THE CLIMATE MAJORITY

About the author

Leo Barasi is a climate and energy policy analyst and an expert in public opinion and campaigns. He has conducted polling and focus groups for political parties, governments and companies across the world. As a campaign consultant he has worked in sectors as diverse as the environment, wealth and health inequality, international development, drug addiction and access to the arts. He lives in London, UK.

Acknowledgements

Many people have made *The Climate Majority* possible. It is built on the research of social scientists and climate scientists – I cannot thank them individually but the book would not exist without their work.

I'm grateful to a number of people who gave their time to discuss ideas and answer my questions. They include: Dena Barasi, Alice Bell, Ann Clark, Jamie Clarke, Tom Crompton, Ruth Davis, Simon Evans, Laura Forrest Smith, Una Galani, Leo Hickman, Chris Hope, Neil Hughes, Ed King, Mark Lynas, Laura Mackinnon, Ben Ormerod, Keiran Pedley, Andrew Pendleton, Roz Pidcock, Stephan Price, Joeri Rogelj, Chris Rose, Kerry Saretsky, Guy Shrubsole, George Smith, Fran Stephenson, Sophie Yeo and the people who helped me understand their opinions on climate change.

I would particularly like to thank a few people who went to an unreasonable amount of trouble to help develop and challenge my ideas: Fred Barasi, Stephen Barasi, Patrick Griffiths, Daniel Harris, Christian Hunt, Rob Vance, Robin Webster and Deborah Wilson. Any errors or omissions that remain are my own.

I would also like to thank the team at New Internationalist for helping to shape the book and for taking it out into the world.

And, above all, thank you to my wife, Susie. Without her patience, support and advice I would never have written the book.

THE CLIMATE MAJORITY

Apathy and action in an age of nationalism

Leo Barasi

The Climate Majority: Apathy and action in an age of nationalism

First published in 2017 by
New Internationalist Publications Ltd
The Old Music Hall
106-108 Cowley Road
Oxford OX4 1JE, UK
newint.org

Editor: Jo Lateu
Design: Andrew Kokotka

Printed by Bell and Bain Ltd, Glasgow.
who hold environmental accreditation ISO 14001.

British Library Cataloguing-in-Publication Data
A catalogue record for this book is available from the British Library.
Library of Congress Cataloging-in-Publication Data
A catalog record for this book is available from the Library of Congress.

ISBN 978-1-78026-407-3
(ISBN ebook 978-1-78026-408-0)

Contents

Introduction

The world has begun to admit it has a temperature problem. Where once the relentless heating of the planet was mentioned only in scientists' charts and protesters' chants, it is now in the mainstream. Newspapers report it on their front pages, businesses plan for a hotter future and world leaders pledge to cut their countries' emissions. When a politician denies the reality of the threat, the global response is now bafflement and ridicule.

But admitting a problem is not the same as dealing with it. Although the world has grown comfortable with talking about climate change, it almost never confronts what it will need to do to cut emissions. Preventing extreme climate change is one of the hardest tasks humans have ever faced. It won't be difficult for just an unfortunate few. Disastrous warming is close to inevitable unless many people, particularly those in rich countries, are prepared to accept sacrifices. Yet, inescapable though this may be, it could barely be further from the minds of the people whose lives would need to change. The world is in the jaws of a trap and has almost no idea how hard it will be to escape.

It is not easy to see when something is missing, but occasionally someone reveals the space where a conversation should be. This happened in 2016 when the Twitter account of an agricultural arm of Bayer, a German company, posted a link to an article that said 'going vegetarian can cut your food carbon footprint in half'. Bayer's social-media manager presumably thought the post would show the company to be taking part in a difficult debate about the future of food. But the account's followers didn't see it that way. The tweet was met with outrage from farmers, appalled that their supplier appeared to be encouraging the public to eat less meat.

Fury on social media is hardly unusual, but what was telling about this outburst was how Bayer responded to the threat of customer boycotts, which was sparked by them sharing an inconvenient truth. They could have stood by the evidence, acknowledged the scale of the challenge and emphasized how they would help farmers prosper without contributing to climate change.[1] Instead, they grovelled. The company deleted the tweet, claiming it hadn't represented their views and that they really did love meat.[2] The incident was left there, a marker on the boundary of what it is acceptable to talk about.

This absence of public recognition of what it will take to deal with climate change may be the greatest challenge now facing the world. The fact that global warming is real and requires attention is no longer unspoken – nearly every country in the world has now declared it will limit its greenhouse-gas emissions. But on the question of what the world needs to do to deal with the problem there is almost complete silence. This failure to confront the reality of what must be done to prevent climate change getting out of control means the world is hurtling towards disaster.

If the world knows that climate change is a problem and has accepted that it needs to deal with it, why is it so reluctant to cut its emissions anything like as quickly as needed?

Some people would answer this by pointing to the influence of oil, gas and coal money on politics and the desire of the powerful to avoid disturbing an economic model that works well for them. They might add that electoral cycles and the length of most political careers encourage short-term thinking. Others would say the difficulty lies in rapidly transforming a global system of energy supply that has been in place for generations – however committed everyone may be to the principle of cutting emissions, the inertia is hard to overcome.

All of these explanations deserve to be explored in detail, but this book looks at another factor behind the world's foot-dragging: the influence of public opinion. It assumes that governments of most high-emitting democratic countries are, at least in part, responsive to what the public wants them to do.[3] If a government thinks the voters whose support it needs are strongly in favor of radical action to cut emissions, such policies will be more likely to follow. But if it thinks most of its supporters aren't persuaded by such measures – or are even hostile to them – then it will be more reluctant to act. This isn't to say that public opinion is static and that it is the only factor behind what a government does about problems like climate change. But, given that effective measures to tackle climate change would inevitably require significant adjustments to most people's lives, public support is vital. Without widespread backing for the measures that will be needed, it is hard to see the world avoiding disastrous global warming.

So this book explores why so many people seem to care so little about climate change and considers what would persuade a majority to support effective measures to address it. If the world can find ways to foster visible public appetite for what is needed to avert extreme warming, then, all being equal, we should expect firmer steps to limit emissions.

But the book's focus is not on the people who do most of the talking about climate change. On one side, the people who reject the consistent findings of climate scientists are far less widespread than they seem and get far more attention than they deserve. On the other side are climate activists. The world owes an enormous debt to the small band of people who have forced governments and the worst emitters to begin to act to limit warming. But, however brave and vocal these protesters are, their actions can never be enough. As long as worries about global warming seem to be restricted to a small minority, governments and businesses won't be convinced that they should take the kind of decisions that have costs to the public and few upsides beyond cutting emissions. For that to change, there needs to be widespread popular concern.

Instead, the problem that most puts the world at risk of disaster is climate apathy. While climate deniers furiously denounce what they see as a conspiracy between scientists, the UN and Al Gore to make us all pay more taxes to a tyrannical world government, the people whose opinions matter most don't care that much either way. They are not angry at climate scientists, they don't think climate change is a hoax, they don't post comments on websites insisting that carbon dioxide is, in fact, good for us. They just don't think about it very much.

This climate apathy is responsible for the filter that stops the world talking about what it will take to avoid extreme warming. As long as so many people aren't particularly bothered about climate change, there is little appetite for conversations about the difficult measures that are needed to deal with it. Instead, discussions about the issue mostly stick to uncontroversial statements that the world must tackle the problem, or focus only on the easy measures that will never be enough.

Apathy about climate change is a greater problem at a time when there is a trend towards isolationism and closed borders. Reducing emissions depends on international co-operation and trust. But the victories of nationalist candidates, such as Donald Trump, and causes, like the UK's vote to leave the European Union, reflect a growing desire in some countries to reduce international commitments. This nationalism is not the same as climate apathy, but it is likely to weaken efforts to deal with climate change – and widespread apathy means there may be little resistance to that effect.

Campaigns that try to influence public debates sometimes use an approach known as the political model.[4] At its core is a principle that seems so obvious that it might appear barely worth mentioning, but is in fact often overlooked: *a campaign should focus on the people it needs to win over and can most easily persuade*. In other words, if you separate everyone you might want to influence into groups of people who are already on your side, people who are firmly against you, and people who are undecided, then it is the last of these who are likely to warrant the most of your attention. This is the guiding principle of many election campaigns, which prioritize winning over the 'swing voters' they need for success.

Debates about climate change mostly ignore this. On the one hand, far more attention than is justified is given to those who deny the findings of climate science. On the other, many of those who are deeply worried about climate change often act as if everyone, except perhaps climate deniers, shares their motivations. They use arguments that might persuade people like themselves, but they forget that the many people in the middle – who may have heard plenty about climate change but are still not that interested – have different values and interests.

This book applies the political model to climate change. It investigates who climate change's swing voters are, what they think, why they think it, and what might change their mind. It looks at how a focus on these swing voters can help with the task of building a large-enough coalition in support of whatever will be needed to avoid disastrous global warming.

The political model has a bad reputation. Many people consider it responsible for the current alienation from politics, blaming it for producing a series of politicians who prioritize chasing swing voters over following their principles. But this criticism shouldn't blind us to where the political model can be useful. It is a powerful tool for understanding public opinion and how attitudes can be changed. How campaigns use the model is up to them.

Besides, when it comes to climate change, anyone seriously worried about the problem doesn't have much space to compromise on their goals. If the world is to avoid dangerous climate change, emissions must be cut radically.[5] There is no way of fudging this. The purpose of applying the political model to climate change is to understand how to win public backing for whatever might be needed to cut emissions. That means winning over people who aren't yet convinced, which will be easier if these efforts are based on a thorough understanding of their motivations.

Nevertheless, the use of the political model does lead to some uncomfortable conclusions about what may be needed to avoid disaster. At times, the best approach may require using arguments that might not come naturally and abandoning other arguments that may be more comfortable for many environmentalists. This sometimes means being single-minded in the pursuit of winning support for emission cuts and treating other problems as a lower priority. Some readers might disagree with that approach – climate change isn't the only challenge facing the world and you may prefer to prioritize other issues. But climate change seems to present such a threat, and the world's decisions over the next few years will determine its extent so greatly, that it seems reasonable to regard it as one of the world's top priorities. Failing to prevent extreme climate change would exacerbate many other problems; getting it right now would avoid harm to billions of lives over the coming decades and centuries.

The book is divided into three parts. The first part is about 'what' and 'why'. It describes what climate apathy is and why it is different from, and more important than, climate denial. It estimates how many people are apathetic about climate change and looks at what they have in common, and it shows why climate apathy matters.

The second part investigates the causes of climate apathy. It looks at the relationship between human psychology and global warming, and how this makes the world less likely to worry about the threat. Then it investigates the ways in which the consequences of climate change are described, including by people who want to draw attention to it, and how these often make it seem unthreatening.

The final part shows how to combat apathy. It describes how we can demonstrate that climate change is important and relevant to more people, and how we can foster hope about the world's prospects for tackling the problem and confidence that emission-cutting measures are fair.

Most of the data and examples in the book are from the four English-speaking countries with the highest emissions and the most vocal debates about whether it is worth tackling climate change: Australia, Canada, the UK and the US.[6] These countries together contribute around 18 per cent of world emissions of carbon dioxide[7] – a fair proportion, although one that is falling as emissions from other countries grow (in 2004, the figure was 26 per cent). Success, or failure, in cutting emissions in richer countries would not alone make the difference between disaster and safety. But progress in those countries would influence how much effort rapidly growing middle-income countries such as China and India put into cutting their own emissions. Although China is now the world's top investor in clean energy and is increasingly pushing other countries to go further in limiting warming, if richer countries failed to cut their own emissions, China's commitment would surely be weakened. But even if the success of richer countries in cutting emissions had no influence on other countries, it would still be the right thing to do in a world where uncontrolled climate change could be an existential threat. And while the focus is mostly on these four English-speaking countries, climate apathy isn't confined to them, and the conclusions about how to tackle it also apply elsewhere.

Since you have picked up a book about climate change – one that doesn't have words like 'fraud', 'hoax' or 'conspiracy' in the title – I assume that you accept that emissions present a serious threat. As such, I use the pronoun 'we' from time to time. Some people have criticized the use of 'we' in the context of climate change, on the basis that it is not always clear whom is being referred to and it risks reinforcing divisions between those who are worried about the problem and those who aren't.[8] This seems fair and I use 'we' only in a specific sense, referring to people who are more worried than average about climate change and are concerned about the slowness of global action to address it. This isn't to suggest that 'we' are the only people who will need to act to deal with the problem, but simply that our existing concern about climate change means 'we' are the ones most likely to be willing to persuade others that it matters.

If we better understand the causes of climate apathy and how we can shift public opinion, we will have a chance of transforming how the world responds to the threat of global warming. Currently, billions of people around the world are apathetic about the issue. Until that apathy is overcome, it will be as if the world is driving with the brakes on as it tries to escape the disaster that is rapidly closing in. Turning apathy into support for serious measures to cut emissions would give everyone a much better chance of reaching safety.

Part 1
Why climate apathy matters

1

The wrong target

Few people think climate change is a hoax. Instead of feeding a fringe conspiracy theory, we should focus on the people who can be persuaded to support the measures that will be needed to prevent disastrous global warming.

'Do not swallow bait offered by the enemy.'
– Sun Tzu, sixth-century BCE Chinese general[1]

To people worried about global warming, climate denial can seem a monstrous foe, but in reality it isn't so mighty. Public doubts about global warming are widely overstated, deliberately exaggerated by people who want to delay action. As a result, efforts to address climate change have become sidetracked by a conspiracy theory that deserves far less attention than it gets. Since climate denial is so established in the debate about global warming, this claim that it is actually unimportant will need some justification. This chapter outlines the evidence for its triviality, arguing that it is far from widespread and that people worried about global warming should learn from successful campaigns, which concentrate their efforts where they can be most effective.

First: what is climate denial? There are different, sometimes contradictory, views that could be labelled 'denial', but they are united by a conviction that current and impending climate change is nothing to worry about. The justifications for this belief include that the climate isn't changing, that any changes aren't caused by human activities, and that, while climate change is real and caused by humans, it won't have much impact. Some people flit between different positions, at times saying they think climate change doesn't exist, while at other times agreeing it is real but saying it won't make much difference. While those who hold these views tend to describe themselves as 'skeptics', the strength of the evidence against them is such that the term 'deniers' is more accurate.

There are other reasons some people give for arguing that the world shouldn't worry about climate change. Some say that long-term economic growth will ensure those affected can afford to deal with it in the future. Others anticipate that revolutionary technological fixes will stop global warming. Similarly, some accept that global warming is likely to be a problem but say people in richer countries should prioritize adapting to its impact at home rather than helping poorer countries do the same. 'Motivated reasoning' is a factor – some people who dismiss climate science seem to do so because of an unconscious desire to maintain their existing beliefs; for example, that less regulation is always preferable.[2] This motivated reasoning might explain why nationalists and believers in unregulated markets are often the most certain that they have found flaws in climate science. They might honestly believe that they have reviewed the evidence about climate change with an open mind, but in fact their attachment to their initial assumptions may have prevented them reaching an uncomfortable conclusion. This can mean it is impossible to distinguish between an honest view that there is likely to be a slightly lower level of warming than the majority of scientists think, and a doctrinaire claim that warming will never be important.

But it is still possible to come up with a usable definition of climate denial: it is the view that global warming won't be a problem and so doesn't require significant attention. This isn't perfect, as it might capture a few people who have arrived at that conclusion with a genuinely open mind, and perhaps would be better described as climate skeptics. Since denial is (intentionally) a pejorative term, grouping all of these people together might seem unfair. But this book is about the people who can still be persuaded to support action to avoid dangerous global warming. Since everyone who meets this definition of climate denial should consistently oppose such action, it is enough for our purposes to consider them together, rather than to devote more time to distinctions between them.

Climate denial is a constant feature of debates about the issue. Sometimes that's explicit – most debates about what the public think of climate change, and whether they are willing to take action, focus on whether they believe in it. When a poll of public opinion about climate change is published, the questions that get the most attention are usually those about whether the climate is changing and whether such changes are caused by humans. If a survey finds that agreement has gone up or down, the results usually become a news story.

Denial also features in discussion of climate change in ways that go beyond analysis of public opinion. Most public-service broadcasters have requirements to reflect a diversity of opinion in their news programs,[3] and they have often interpreted this to mean that they should include the views of climate deniers in coverage of global warming.

A debate on the BBC's top current affairs radio program, *Today*, exemplified the problem. In February 2014, Professor Brian Hoskins, an eminent climate scientist, was interviewed alongside Lord Lawson of Blaby, a Conservative politician who repeated throughout the eight-minute interview that 'nobody knows' whether there is a link between climate change and extreme weather. Eventually, the presenter, Justin Webb, a senior journalist who normally shows no fear of politicians, asked the peer a question whose timidity makes it worth quoting in full:

> 'Can I just put this to you though, if there is a chance – and some people would say there is a strong chance – that man-made global warming exists, and is having an impact on us, doesn't it make sense, whether or not you believe that that is a 95-per-cent chance or a 50-per-cent chance or whatever, doesn't it make sense to take care to try to avoid the kind of emissions that may be contributing to it?'[4]

With the debate presented as being between two uncertain perspectives, Brian Hoskins tried to focus on the evidence, but with little success:

> *Scientist:* The excess energy is still being absorbed by the climate system, and it's being absorbed by the ocean, so the ocean's warming up.
>
> *Politician:* That is pure speculation.
>
> *Scientist:* No it's measurement.
>
> *Politician:* No it's not, it's speculation, with great respect.

Presenter: Well it's a combination of the two, isn't it, as is this whole discussion.

The debate ended there, with listeners told that the true answer was somewhere between the extremes.

A BBC internal investigation upheld a complaint against that interview, finding that it was a mistake to have treated the politician's opinion as being on the same footing as the scientific evidence,[5] echoing another internal report that had said the broadcaster gave 'undue attention to marginal opinion'.[6] Yet since then, *Today* has continued to give prime interview slots to people who say that climate change doesn't represent a significant threat.[7] The Australian[8] and Canadian[9] public broadcasters have been similarly criticized for providing prominent space to interviewees who reject the conclusions of the overwhelming majority of climate scientists.

Many newspapers – and some broadcasters that have no requirement for balance, particularly those in the US – go further and assert that climate change is exaggerated or a hoax, without even presenting the opposing argument. A study of US and UK coverage of the 2013 report of the UN's climate science body, the Intergovernmental Panel on Climate Change (IPCC), found that it over-represented doubts about climate science, by a factor of five in the case of the US.[10]

The same is true outside the mainstream media. News articles or blogposts about any aspect of climate change frequently attract comments arguing that it is a hoax, or at least greatly exaggerated.[11] These seem easily to outnumber the responses arguing the opposite, giving the impression that climate denial is widespread among the public.

Denial is common in the politics of some countries, too. The US may be unusual in having, in Donald Trump, a president who has called climate change a hoax, but past leaders of both Australia and Canada have fought against measures that would reduce emissions.[12] Among these countries, the UK has generally been an exception – so far, its Conservative governments have expressed less doubt about climate change than their counterparts in other major English-speaking countries (this is also the case in some other Anglophone countries, such as New Zealand/Aotearoa).

This denial is so pervasive that it is reflected even in the words of people who have no serious doubts about the causes and likely severity of climate change. Open any major report on climate change and there's a good chance that one of the first things you will see is a defensive response to denial. President Obama's Climate Action Plan started with the words: 'Some may still deny the overwhelming judgement of science.'[13] The first report of the UK's Adaptation Sub-Committee began: 'The overwhelming majority of experts agree that the global climate is changing, and that most of this is caused by human activity.'[14] Even the Pope's climate encyclical said: 'A number of scientific studies indicate that most global warming in recent decades is due to... human activity.'[15] These references to scientific evidence about the causes of climate change, in reports about dealing with the problem, show that denial is influential not only in the way it generates repetitive debates about whether the public thinks climate change is real, but also in how it distracts those who have no such doubts.

We rarely notice the extent of this influence, but we should. A study of around 12,000 peer-reviewed papers on climate science found that 97 per cent of those that stated a position agreed that current global warming is caused by humans,[16] while the IPCC's reports show that the impacts of unrestricted climate change would be disastrous for billions around the world.[17] Still, most political, public and media discussions about climate change take place as if there's an ongoing argument about whether it is real, caused by humans, or likely to be a problem.

But despite the reach of climate denial, successful campaigns in all fields show us that there are better uses of time, money and energy than focusing on the arguments of people who steadfastly oppose our goal.

Campaigns and audiences

In December 1994, the US's Democratic president Bill Clinton was in trouble. The Republicans had just taken control of both houses of Congress for the first time since 1952 and in less than two years he would face re-election. His popularity was falling rapidly, with only 40 per cent approving of his performance as president. Such a low rating wasn't encouraging – the previous Democratic incumbent,

Jimmy Carter, had an approval rating 10 percentage points higher at the same stage and was still thrown out by the voters.

Attacks on the president were coming from Democrats as well as Republicans. In the recent midterm elections, the Republicans had fought with the promise of a series of reforms, set out in their 'Contract with America', and they now considered themselves to have a mandate for change. Urging them on was an army of activists, furious at what they saw as social liberalism's overreach. Their leaders included Pat Buchanan, a former Nixon and Reagan staffer, who declared that he was fighting a 'cultural war... for the soul of America'. His targets included the 'radical feminist' Hillary Clinton, whom he accused of wanting to impose, among other things, 'homosexual rights'.[18] And pressing Bill Clinton from his own side was the Democrats' host of activists, furious at the Republicans and angry at the President for not seeming to be fighting back against the conservative onslaught.

But, satisfying though engaging in a culture war might have been for the President and his aides, it couldn't assure him of re-election. In the previous decade there had been a fundamental shift in US political identities – for the first time, more people identified as Independents than as Democrats.[19] Even if Clinton could rally everyone who considered themselves to be in his camp, he could still be overwhelmed by a Republican opponent that won over more Independents. The only way for Clinton not to be a one-term president was to win the votes of people who didn't call themselves Democrats.

This became the focus of his next 18 months. Clinton didn't shy away from fighting the Republicans, but he chose the battlefields that he knew Independents cared about – the economy, jobs and public safety – rather than those his committed supporters prioritized. In 1995 he allowed a confrontation with congressional Republicans to shut down the government, correctly calculating that his opponents would be blamed for letting their ideology harm ordinary people. With a relentless focus on the issues that mattered most to Independents – which helped him draw attention to the growing economy – Clinton's approval rating touched 60 per cent in 1996 and he was re-elected with a margin of victory that remains the largest since Ronald Reagan's in 1984.

Every campaign, everywhere in the world, has limited money and staff time. At one end of the scale, the presidential campaigns that seek to emulate Bill Clinton's successes now spend over $1 billion and have enormous numbers of volunteers. At the other, local voluntary campaigns across the world rely on the spare time and money of just a handful of people. But, whichever end we're looking at, a campaign's success depends on it using its assets effectively.

Campaigns seek to use whatever time and money they have to influence what people think, and this first requires getting their attention. However rich a campaign may be, this is always a challenge. One campaign strategist has claimed that the average person spends no more than four minutes a week thinking about politics, even in the run-up to an election.[20] Whether or not that figure is based on much evidence, it reflects an important point. Most people, most of the time, don't pay much attention to political – or other – campaigns. Any attention a campaign receives is potentially valuable.

So, the second task for any campaign is making sure it uses that attention to its advantage. What the campaign says and does needs to influence the people it is talking to. To plan a route to victory, a campaign needs to understand its audiences. It needs to know how likely different people are to support its goals and what might change their minds. Without this knowledge, a campaign struggles to make the most of whatever attention it can secure.

The public aren't a homogeneous block with the same interests and motivations, and a good campaign doesn't treat them as such. Effective campaigns recognize the diversity of views in their audiences and plan their activities accordingly. To do this, they divide the public[21] into notional groups. Members of each group should share enough for a campaign to be able to influence them with common approaches. This process is known as segmentation.

There are many ways of creating segmentations. Sometimes individuals are categorized on the basis of one or two habits, such as whether or not they support the goals of the campaign or are customers of the business. Other campaigns use demographic factors, such as age, gender, ethnicity, home ownership and whether or not someone has children. This demographic approach is the one most often discussed in political commentary – for his re-election Bill Clinton was widely reported to be targeting busy middle-income women with children, known as 'Soccer Moms'. While such approaches are an improvement on treating the public as a single group, a more subtle methodology is often better still.

Instead of segmenting on the basis of demographics or just one or two characteristics, some campaigns draw on a wider range of attributes. This approach uses statistical processes known as cluster analysis, which draw on large datasets to identify groups of people – segments – who share particular characteristics. These characteristics are usually not demographic, but more often focus on attitudes and behaviors that the campaign is seeking to change. Using these specific attitudes distinguishes this approach to segmentation from analysis that is based entirely on values, which some campaigns use, although the segments produced may be similar.[22] The analysis seeks to define groups whose interests coincide sufficiently for it to be possible to develop policies, actions or arguments that most members of a particular group will find appealing. This is much more achievable with a segment that, for example, is defined by their worries about inequality and housing than it is with one that is defined as, for example, women aged between 30 and 49. That said, since they share attitudes and behaviors, members of a segment do also tend to share particular demographic characteristics – but this is an output of the process, rather than an input.[23]

This use of attitudes and behaviors, rather than demographics or values, as the basis of the segmentation means there isn't a universal segmentation of this kind that can be used in every campaign. Analysis carried out this way organizes people according to the factors that a particular campaign most wants to change. This might be willingness to act on climate change or likelihood to buy a category of smartphone. That makes the process more expensive for each campaign – as the work has to be done afresh each time – but it also makes the results more useful.

This approach to segmentation is used widely in the private sector. When a major company wants to develop and target its products – whether fast food, smartphones or TV packages – it often develops a segmentation to guide its decision. After detailed research, drawing on buying habits, questionnaires, focus groups and watching people go about their daily lives, the company develops detailed descriptions of its existing and potential markets. It estimates the size of the segments and tries to understand what motivates people in each of them – what do they want from the products and how can the company meet those desires? It is a complicated and lengthy process but it can create a powerful way of understanding what different people think, to help develop a campaign, policy or product that appeals to specific groups, rather than a lowest-common-denominator one that nobody objects to, but which doesn't really excite anybody.

Some campaigns go further and use more detailed data to create ever-smaller groups among the population. The best-resourced campaigns have been doing this for years, but Facebook has made it possible for more campaigns to 'micro-target' audiences in groups that are effectively as small as they choose. It has been suggested that this was one reason for Donald Trump's unexpected election victory in 2016.[24] The availability of such detailed data is likely to mean that an increasing number of campaigns will move away from targeting a few large segments, and will focus instead on more nuanced targeting of many small segments. This undoubtedly allows campaigns to focus their messages as effectively as possible – but it is hard to think about the larger trends if you are constantly thinking about tiny sub-groups. So, for the purposes of this book, we will remain at the level of a few large segments. The differences between these segments are large enough for us to see how we can win over many more people to caring about climate change, without needing to get down to the level of micro-targeting.

The most useful aspect of segmentation is that it helps a campaign to prioritize. After segments have been defined, each can be ranked according to where they stand on the issue that the campaign wants to influence. At one end of the scale are the segments that are the strongest supporters. These are sometimes described as the 'base'. At the other end are those that oppose the campaign's goals. In politics they are usually known as 'critics'.[25] In between are the 'swings'[26] (as in, swing voters), who aren't yet supporters or customers, but might plausibly be persuaded by either side. Within each of these broad groups there is often more than one segment.[27]

Campaigns treat these broad groups in different ways. In general, people in the base segments are often already supporters or customers, so the objective with this group is to win over any who are not yet supporters, retain the backing of others and, if possible, encourage them to become advocates. These are the people who might talk to their friends or family, show their support on social media, and knock on doors or donate during elections. In some contexts they may also become more involved with the campaign, for example by bringing their expertise to help shape policy.

The goal with the swings is to win them over – to move them from being undecided or weak supporters of either the cause or its rivals, to being committed supporters of the campaign. The segmentation should be designed to help campaign planners understand how they can do this.

This leaves the critics. They can't always be ignored – if unchallenged, their arguments might influence swings and even some people in the base segments. What they are saying might need to be addressed, although sometimes their criticisms are better ignored rather than dignified. But winning or persuading these segments isn't a priority, and most well-targeted campaigns decide it is harder to win over people in these distant segments than it is to persuade those in the swing groups.

Bill Clinton's focus on Independents for his re-election reflects this, and plenty of other campaigns have taken a similar approach. When politicians and campaigners have argued in favor of same-sex marriage, they have often faced opposition from people who consider it an unacceptable redefinition of a traditional institution, with some saying it is against religious teaching. Many of these opponents are never going to be won over to support equal marriage in the timeframe of a referendum or legislative process, but their arguments could still influence others and need to be addressed. Similarly, when campaigns have fought to restrict public smoking, they have usually been met with furious opposition from a small group of passionate smoking advocates and libertarians. But these anti-smoking campaigns have largely made progress because they have recognized that their primary goal isn't to win over those opponents, it is to persuade the larger group in the middle that tighter controls are right for most people.

The experience of such campaigns is that victory doesn't depend on winning the support of opponents, even if winning the swings sometimes depends on overcoming these opponents' arguments.

When it comes to limiting climate change, the critics are those who reject the view that there is a problem that needs to be addressed. So if climate denial really were widespread, we would have a problem, since our objective is to build clear support for measures to cut emissions. If deniers made up such a large proportion of the population that widespread support was impossible without their backing, we couldn't treat them in the same way that campaigners for equal marriage or for restrictions on public smoking treated their critics: as opponents, rather than potential converts. The extent of climate denial is therefore an important question.

Exaggerated doubts

Opinion polls show how many people fit this chapter's definition of climate denial. It is true that no single question provides a full picture of an individual's views and it is better to use a series of measures – this is what the next chapter does as it seeks to understand climate apathy. But before we get there we need first to be confident that climate denial really is rare. While individual poll questions might not tell us everything we need to know about public opinion, they can put denial into context.

Across major industrialized Anglophone countries, typically less than 15 per cent answer surveys with views that would meet this chapter's definition of denial. The proportions vary depending on the wording of the question, but it is rare for a poll to find more than a few people strongly rejecting climate scientists' conclusions about the causes and likely consequences of climate change. For example, in a 2015 international poll, 10 per cent in Australia, 5 per cent in Canada, 7 per cent in the UK and 12 per cent in the US said they think global climate change is not a problem.[28]

In most rich countries, concern about climate change appears to have fallen in the years up to 2010, and to have increased a little since then. Even during the years in which concern fell, there was only a small increase in the numbers saying they thought it was a hoax. Most of the change was in the proportion that moved from being convinced about the problem to being less sure about it. For example, in the UK, the proportion who told a repeated survey that they were not at all concerned about climate change increased from 3 to 8 per cent, while the proportion that said they were in the middle – either fairly concerned or not very concerned – jumped from 50 per cent to 62 per cent.[29] A similar trend occurred in both Australia and Canada – up to around 2010 there was a large increase in the proportion who said they were unsure that climate change was caused by human activity, with only a small increase in those saying they didn't think it was happening at all. This appears to have largely been reversed since then.[30] In the US, concern about global warming recovered by 2016 to around the level it was at in 2008.[31] Trends in other parts of the world have been less consistent, with some countries showing increasing concern over the last decade and others showing a decline.[32]

To put this in context, we can compare the extent of climate denial with what other polls have found about opinions that the mainstream media and top politicians don't take so seriously. A poll in Australia suggested that 22 per cent believe in witches, while another in Canada found that 21 per cent said they think Bigfoot is real[33] – in both cases more than twice as many as those who say climate change won't be a threat. A 2013 poll found that 33 per cent in the UK believe that Princess Diana was assassinated, not killed accidentally[34] – more than four times as many as those who say they don't think climate change will be a threat. Even in the country regarded as most doubtful of climate change, the US, 24 per cent said they think the government knew about the 9/11 attacks in advance but did nothing to stop them,[35] twice as many as believe climate change won't be a problem.

So it seems fair to say that only a small part of the public deny that climate change is a threat. We don't treat the fact that up to a third of people believe conspiracy theories – such as that Princess Diana was murdered, or that 9/11 was an inside job, or that Bigfoot and witches prowl the land – as evidence that these beliefs are common. This seems encouraging for the question of whether widespread support for measures to address climate change depends on people who currently think it is a hoax.

Another reason to think we have been worrying too much about denial is that the strategy of those who want to stop action on climate change is to overstate doubts about it. In a 2002 memo, a political consultant, Frank Luntz, told President George W Bush that the Republican Party was in danger of losing the argument. At the top of Luntz's list of what Republicans should do to avoid defeat was a warning that, 'should the public come to believe that the scientific issues are settled, their views about global warming will change accordingly'.[36] The purpose, in Luntz's view, of overstating doubts about climate change was not to refute the conclusions of climate science – he didn't seem to think that was possible. Instead, he suggested Republicans should create enough doubt about climate science to delay action. He argued that 'there is still a window of opportunity to challenge the science'.

We shouldn't overstate the importance of this particular memo – it was just the view of one consultant to the US Republican Party, which no doubt also heard many other views. But regardless of whether Luntz personally shaped the strategy of those opposing action on climate change, his approach has been adopted. As Naomi Oreskes and Erik M Conway have shown in their book, *Merchants of Doubt*, the tactic of casting doubt on science to delay action has been used by opponents of emissions control since at least 1989, drawing on the earlier successes of tobacco companies.[37] The delayers' purpose isn't to win the argument, it is to maintain the impression that there is still an argument.

Given this, it is significant that climate denial features so prominently in the media. If most of the public don't believe that climate change is a hoax, and vanishingly few relevant scientists think it is, why should denial command so much media interest? An obvious answer is that it is because there is a vocal group of organizations, some of which are funded by fossil-fuel companies,[38] that provide reports and articulate interviewees arguing that climate change isn't an important problem. But why should their arguments be taken seriously?

One explanation may be that the media seek to appeal to their audiences' existing opinions, and, while climate denial is not particularly widespread, its adherents tend to read or watch the same outlets. There used to be just a few places where someone could go for news – a newspaper or major TV or radio channel –

but there are now effectively unlimited TV channels and websites. It is becoming ever easier for someone to find news that confirms their existing beliefs – a trend that is reinforced by social media and the proliferation of climate-denying blogs.[39] But, if this was the only explanation, we should expect denial to be restricted to just a small collection of news outlets, serving around 15 per cent of the public, with a much larger number of providers accepting the findings of climate scientists. Since this isn't the case, there must be other explanations.

A second possible factor is that journalists often want to challenge accepted wisdom. Climate change is widely accepted by governments, international institutions and campaign groups, so it seems ripe for attack by those who want to tear down orthodoxies. What's more, a journalist might think, science isn't a collection of facts – it is a process of developing ideas, and that process advances through ideas being challenged. So the media shouldn't go to the extreme of never covering stories that question scientific orthodoxies, in case they miss the next Galileo. But this can be wildly over-interpreted. Modern scientists understand the planet's climate system much better than Galileo's contemporaries understood the motion of heavenly bodies. Just because some orthodoxies have been overturned doesn't mean journalists should treat every collection of scientific knowledge as if it is no more valid than any other opinion.

A third explanation is that attacking climate science gives journalists something dramatic to cover when climate change otherwise seems not to generate much news to interest a non-specialist audience. Regardless of the strength of the evidence, arguments about whether or not climate change is real and whether scientists are falsifying data can offer the tension that journalists and editors look for in news stories.

These might not seem like good reasons for giving so much space to climate denial. We don't expect the same in media coverage of, for example, anti-smoking policy or research into particle physics. But this is the reality of how the media treats climate change, and those of us worried about the issue need to find ways of dealing with it, rather than just bemoaning it.

So it doesn't seem realistic to ignore climate denial entirely. It has become too widely established as a part of the debate, even if only a small part of the public fully accept it, and the organizations that oppose cutting emissions aren't about to stop promoting misleading information about climate science. It seems inescapable that we must address these arguments, or else they will continue to dominate discussions of climate change at the expense of more productive debates. Yet this seems to play into the hands of those, like Luntz, who want to focus the conversation on supposed doubts about climate science. Years of responding to denial with scientific evidence have not dispelled the perception of doubts.

The solution may be to change the subject. When the subject of denial is raised, people worried about climate change should firmly reject – even ridicule – the suggestion that the basic science is still unsettled and then move the conversation on. Dealing robustly with climate denial is necessary, or it will continue to be treated as if it is widespread and taken seriously by climate scientists. The same is true with the organizations that promote climate denial – exposing their links with fossil-fuel interests, their history of scientific errors and the motivated reasoning that consistently leads them to conclusions that reflect their anti-regulation ideology can undermine their claim to be impartial and reliable participants in the debate. But there is little benefit in perpetuating the way these misrepresentations of science and public opinion are the focus of debates about global warming. As in the BBC radio debate described above, attempting to deal with each individual inaccuracy can be like trying to catch fog in a net. In any debate, deniers can misrepresent the evidence, refer to irrelevant or unrepresentative facts or just change the subject. Audiences that hear about an issue for the first time this way are left no closer to knowing what climate scientists really think. Instead of engaging in this attritional warfare, there are far more productive areas for debate – we should move from talking about whether there is a problem to instead discussing what the world should do about it.

We should be realistic about how far this is likely to end media coverage of denial, and, indeed, whether doing so is even necessary. The incentives for some parts of the media to cater for climate-denying audiences are beyond our control. Since there is now such a wide variety of news sources available, there may always be a selection of news outlets that consistently deny climate change – perhaps the same ones that run stories about ghosts and UFOs. But, as long as these sources cater overwhelmingly to people who already passionately oppose all measures to deal with climate change, that doesn't need to matter. It's more important to shift other parts of the media – those that are seen by the swing audiences – to talk about something else. This will mean tackling the other reasons the media disproportionately cover denial: that it allows them to challenge orthodoxies and to inject drama into the news. Part 2 discusses how we could do this.

Unlike some other campaigns, efforts to persuade more people to back measures to limit climate change don't have a single deadline. In an election, voters make a one-off decision that they are usually stuck with for a few years. Since dealing with climate change is an ongoing problem, the manner in which critics' arguments are addressed is particularly important. Taking part in conversations about denial and using the opportunity to emphasize other, more productive, topics – rather than just refusing to engage – may reduce the risk that climate deniers can present themselves as victims who are shut out of a closed debate. It may not change many opponents' minds, but it allows those of us who are worried about the problem to shift the conversation onto topics that might influence other people.

For now, climate denial continues to attract a disproportionate level of attention. Some of that attention unintentionally comes from people who are seeking to address climate change but who have become distracted arguing with a view that is less widespread than a host of fringe conspiracy theories. All campaigns benefit from focusing resources where they are most effective, but efforts to encourage action on global warming have often failed to do this, with the result that years have been wasted trying to change the minds of people who will never be persuaded. This is exactly what some opponents of measures to cut emissions intend when they seek to exaggerate doubts about it.

Having raised this problem with the attention that is given to climate denial, let's now put it to one side. The rest of the book focuses on the people whose opinion will influence whether or not the world can avoid extreme global warming. These are the swing audiences who are often apathetic about the issue, but whose views are very different from those of the people who think climate change is a hoax. The remainder of Part 1 describes these swings and shows why they will determine whether or not the world avoids disaster.

2

Who cares about climate change?

Around half of the population accept that climate change is happening yet are apathetic about it. Unlike climate deniers, these 'swing voters' could be persuaded that the issue requires urgent action – but at the moment it is not obvious to them why it matters.

'The essence of strategy is choosing what not to do.'
– Michael Porter, professor of management and strategy, Harvard University[1]

The Copenhagen climate change conference in 2009 is generally remembered as a disaster. In the months of build-up, when some campaigners labelled the conference 'Hopenhagen', there was widespread optimism that it would be a turning point in the world's efforts to avoid catastrophic global warming. That was proved correct, but not in the way many campaigners had hoped. The talks collapsed, with a limp statement of aspirations the only thing to cover the hosts' embarrassment.[2] The failure led many people to view the conference as the moment it became clear that national governments would never limit emissions to safe levels.

But for me, there was a small consolation. As a result of the conference, I noticed a statistic that made me realize that much of what I thought about how the public see climate change was wrong. As I learned more, it became clear that public opinion is even more challenging than it had seemed for those worried about the climate. But I also came to appreciate that there are reasons to be optimistic that more people could be won over to support the measures that will be needed.

The statistic was buried in a poll conducted shortly after the climate talks ended, which looked at how people in the UK viewed the conference agreement, the Copenhagen Accord. The poll found overall support for the Accord's goals, which wasn't surprising. But what was unexpected was how the Accord was seen by people who also said they think climate change is a natural phenomenon, rather than one caused by humans. Logically, these doubters shouldn't be keen on action to cut emissions. If climate change isn't caused by human activity, there would be no reason to impose restrictions that might make fuel more expensive, impede heavy industry and limit other aspects of life. But the results didn't follow that logic. Instead of finding resistance to an apparently unnecessary agreement, the poll showed that even most people *who thought climate change was mostly a natural phenomenon* were behind the goal of the Accord to cut emissions. Among these doubters, 69 per cent supported it, while only 24 per cent opposed it.[3]

It was this number – the 69 per cent – that got me interested in the people in the middle. In the arguments about climate change it is easy to overlook those who neither think the world is warming because of human activities, nor believe it is a hoax. They don't fit into the easy categories of environmentalists or deniers. They seem grey when others have views in primary colors. But as I started looking more closely at the people in the middle, it became clear that their views are much more interesting than I had appreciated. They sometimes seem to contradict themselves, with doubts about the causes of climate change coexisting with support for action to address it. And they represent much more of the public than I had imagined – in some places more than half the population. As such a large group, they hold the key to determining whether or not measures to limit climate change can get mass public support. I knew I had to understand their views better: what they really think, why they think it, and what – if anything – could be done to influence them.

These are the swings. The consistent thing about how they see climate change is that they are not consistent. They seem to have the potential to be persuaded by either side of the debate. And when they are considered using the segmentation approach described in the previous chapter, we will see that they are best understood not as one homogeneous block but as a diverse group with different clusters of views. This chapter looks at those different views, to understand the swings and why some of them[4] say they think climate change is natural while also wanting restrictions on emissions. It explores who the swings – as well as the other segments – are and how they can be divided into different groups. This will help us see what motivates them and forms their views of climate change, as a step towards understanding what might persuade them to support action to cut emissions.

Sound and fury

Before we get into the detail of the different segments among the swings, we should first view the landscape from a distance, as there are features of public opinion about climate change that apply across the segments.

Public opinion about climate change is often measured with simple questions about belief that the world is warming and concern about the threat. These questions typically find that between 40 and 60 per cent in Australia, Canada, the UK and US think climate change is real and caused by humans; around 30 to 40 per cent say they think it is natural; and, as we saw in the previous chapter, usually less than 15 per cent say they think it is not happening at all.[5] Similarly, between 25 and 50 per cent typically say climate change is a very serious problem; less than 16 per cent say it is not a serious problem at all; and the rest are in the middle.[6]

It is best not to take those figures literally. They are influenced by the wording of the polls and there is wide variation between surveys. In contrast, segmentation studies draw on a range of questions and give more consistent numbers. They show that there is a larger share of the population in the swing groups than simpler polls suggest. And, as I discovered after the Copenhagen conference, even many of those who appear to doubt climate science still want action to cut emissions. That wasn't some rogue poll – its finding is borne out by other surveys, which consistently find that more people say they want action to limit climate change than say they think it is caused by human activity. For example, a poll conducted ahead of the 2015 Paris climate conference found that 69 per cent of Americans, 80 per cent of Australians, 78 per cent of Britons and 84 per cent of Canadians thought their country should join an international treaty requiring it to reduce its emissions.[7] In all cases, this was more than the percentage who said they thought climate change is real and caused by humans.

But while these studies consistently show majorities in favor of stronger action to limit climate change, it still doesn't seem a pressing concern for most people. When asked about the top issues facing their country or their family, very few people spontaneously mention climate change – typically, no more than 10 per cent, and usually less, do so.[8] This sometimes goes up a little, generally following some weather-related disaster that has been linked to climate change, like floods or fire. Even then it is rare that more than about 25 per cent name climate change as a priority issue, and it quickly falls back to the level it was at previously.[9]

The fact that worries about climate change don't seem to increase much, even after its likely consequences seem evident, might be a sign that most people pay little attention to news about it. That would make it harder to persuade more swings to support measures to address the problem – if they are not interested in news about it, they are less likely to hear stories about the issue.

But is it true that many people don't pay attention to news about climate change? The fact that weather-related disasters don't seem to increase worries about global warming may be a sign of this. Another way to check is to look at the impact of attacks on climate science – that is, to see whether attempts to shift public opinion away from worrying about climate change have had any effect.

Probably the most prominent of these attacks was the publication in November 2009 of emails hacked from a climate research center at the University of East Anglia. Critics claimed that the emails proved that the scientists involved in the research were manipulating data to produce the results they wanted, and labelled the supposed scandal 'Climategate'. The email hack dominated media coverage of climate change ahead of the Copenhagen conference and in the months afterwards, only fading after a series of investigations cleared the scientists accused of fraud. A characteristic of this, and other, attacks has been the personal nature of the criticisms of individual scientists. This has included not only legal challenges and accusations of fakery, but also death threats.[10]

At first sight, these attacks seem to have had some impact on views of climate change. Several UK polls in the weeks after the release of the hacked emails showed a fall in belief that climate change was real and caused by humans, with a corresponding increase in the numbers saying they were unsure about it. It was nearly three years before opinion returned to where it had been before the attack.[11] In the US, trust in climate scientists also fell after the hack.[12]

But it is easy to overstate the impact of these attacks. Horrible though they are for the scientists who are targeted, they don't seem to have undermined public trust in climate science. In fact, climate scientists remain easily the most trusted source on the issue. A 2013 UK poll found 69 per cent consider scientists or meteorologists to be trustworthy.[13] Polls in the US and Canada have similarly shown that far more people say they trust scientists than don't,[14] and scientists remain the most trusted source of information on climate change. Where trust in scientists has been measured over time, there is little evidence of a decline, other than in the US, where trust does seem to have fallen in recent years – although from a high starting point,[15] and principally among those who were already predisposed to doubt climate change.[16] Likewise, concern seems now to have recovered to the level it was at before the 2009 email hack, despite continuing attacks on climate science.

And, while some of the attacks on science do seem to have been followed by greater doubts about climate change – at least for a time – that doesn't mean the 2009 email hack still influences public opinion. It can be tempting to imagine a simple model of how such attacks shape public perceptions: many people hear the criticisms of climate science, weigh up the arguments and become more doubtful. But when your attention is focused on a particular field, it is easy to forget that most other people are usually much less interested in it than you are.

To measure the impact of the email hack, four years after it had been in the news, I worked with the website Carbon Brief[17] to see how many people could remember hearing about it. In a poll, we listed various climate-related stories and asked respondents whether they could remember seeing or hearing each of these.

Since prompting people in this way usually leads to an over-estimation of recall, we also included some fake stories to compare with the real ones.[18] While the story 'scientists have been accused of faking data about climate change' was the second-most recalled, only around one in three of those with doubts about climate change said they could remember hearing it.[19] This wasn't much more than the one in five who thought they could remember the most popular fake story, about China announcing it wouldn't curb its emissions. Even in the US, where opinions on the climate are the most polarized, fewer than one in four people said they could remember hearing about 'Climategate', when asked in a separate study, only a year after the hack.[20] While the hacked emails continue to interest some of those who deny the reality of climate change – as reflected by Donald Trump referring to it in an interview in November 2016[21] – most people forgot about the incident years ago, if they were even aware of it in the first place.

There was another interesting result from that news-recall question. In the years since the email hack, many arguments against cutting emissions have been based on a claim that the world has stopped warming. This idea has been widely criticized – it only appears to be valid with a selective choice of dates or locations, which renders it dubious at best[22] – yet it is often mentioned as a reason not to worry about climate change. But we tested public awareness of this as well and found that only five per cent of people overall, and seven per cent of those who believe climate change is natural, said they could recall hearing a story about climate change having stopped over the past 16 years. This was only slightly more than claimed to recall our silliest fake story, that ocean acidification is making dolphins more intelligent.

So it seems most people generally don't see climate change as a top priority and individual news stories about it are usually either not heard or are quickly forgotten and don't appear directly to influence public concern once they have faded from memory. This isn't to say that such stories have absolutely no impact. It is hard to dispute that ongoing attacks on climate science have increased public doubts about the reality and significance of the problem. Without any such attacks, belief that climate change is real and a problem would, it seems likely, be higher – which is a reason why the continued attention given to climate denial is significant.

One other consequence of the low level of attention paid to climate change is that some misunderstandings about it have been allowed to flourish. In Chapter 6 we will see the ways in which many people mistake the likely consequences of climate change, but there are also other areas of misunderstanding.

Among the most significant of these is that there is widespread misunderstanding about which actions would have the greatest impact on reducing emissions. A UK study found that the top two things most people thought the public can do to reduce emissions were recycling and insulating their homes.[23] The bottom two on their list were eating less meat and dairy, and flying less. The reality is that recycling and improving insulation could make a much smaller contribution to reducing emissions than cutting out meat and flying (the next chapter discusses this in more detail). The exact figures depend on many factors, but making the latter changes could produce a roughly nine-fold greater saving.[24] Another study showed a similar misunderstanding, with a widespread underestimate of the impact of meat and dairy production, and an overestimate of the impact of how waste is treated.[25]

Similarly, many people appear not to have an accurate picture of the efforts they or their country are already making to reduce emissions. In Canada, a 2013 survey found that only one in three people knew that their government had withdrawn from the Kyoto Protocol, an international agreement to limit emissions.[26] Similarly, a surcharge added to household energy bills to fund renewable power was the center of a political argument in the UK in 2013. But after extensive media coverage – much of which exaggerated the cost – most people still couldn't come close to estimating how much they were paying towards renewable power. According to one poll, the average estimate of the subsidy that consumers pay through their bill was £259 ($315), around 14 times higher than the true figure.[27] This isn't to criticize people who make these mistakes – the fact that they don't closely follow the debate is not some moral failing. The point of noting the gap between perception and reality is to show the challenges that we will face in trying to change public attitudes.

So, even before getting into the segments of public opinion, we can already see some important aspects of views of climate change. Many of those who say they are unsure about global warming still want action to avoid it. The numbers of these swings have swelled in recent years, perhaps influenced by the long-running attacks on climate science, although confidence in climate scientists is still high and most people can't remember hearing news about data fakery. Nevertheless, many people misunderstand some key facts about climate change, including what can be done to address it and what their country is already doing about it. Ultimately, most people don't pay much attention to news about global warming. This may have blunted some attacks on climate science, but it also means that the task of persuading more people to support measures to limit warming will be more difficult. Before we can have any chance of winning over the swings, the first challenge will be attracting their attention.

With this overview in mind, let's move in closer and look at how the public can be separated into segments, based on what different people think about climate change.

Meet the swings

Picture a small living room, crowded with 10 people and noisy with bad-tempered argument. As you watch, you realize not everyone is taking part in the argument – in fact less than half of the crowd are talking. One pair, sitting together, is arguing, not with each other, but with another two people, who are firing back retorts to everything the first pair say. And watching the debate, with expressions ranging from bewilderment to boredom, are the remaining six, who so far haven't said anything (in fact, one of them has given up paying attention and is browsing on her phone).

Those quiet six people in the middle are often forgotten, but deserve our attention. In the debate about climate change, they're the swings. If they were to join sides with either of the vocal pairs, the new group would have a comfortable majority. But it is obvious that they are not going to be well disposed to the arguing sides so long as the people doing the talking are ignoring the bystanders. So, who are the swings, what do they think about climate change, and what might influence their views?

Let's start with the swings in the living room and imagine they are representative of the wider public. Three of the six appear a bit more interested in the debate and seem fairly sympathetic to what is being said by the pair arguing that climate change is a serious threat. Later, we will call these three the *Concerned*, and they do indeed seem to be concerned by what they hear about climate change; or at least, they did at first, although after half an hour of argument they seem to have lost interest. If we could ask them what they think, they might say: 'I believe that climate change is happening, it's serious and it should matter, but I think it would be better to prioritize other issues such as housing and the health service.'[28]

The other three swings don't seem so sympathetic to those arguing that we should worry about climate change – sometimes they are also nodding at what they hear from the pair arguing the other way. Mostly, though, they just look a bit bored. We will call these three the *Cautious* – they are not against the idea that climate change is a problem, but they are not convinced by arguments about its urgency or the sacrifices that are proposed. About climate change, they might say: 'I often see targets I don't really understand, with implications I can't really grasp and I'm skeptical when I hear of things such as flight taxes, which I worry won't achieve anything.'[29]

We will now look in more detail at all the segments and, as we do so, it might be useful to keep these two sets of swings in mind. The first are the *Concerned*, the second are the *Cautious*, and neither thinks that climate change is a hoax.

Several segmentation studies have looked at how the public can be divided according to their views on climate change. The studies, in Australia, the UK and US,[30] allow us to see how people split between the base, swing and critic segments, and what those in each group tend to think. Understanding these segments will help us appreciate why there often appears to be so little interest in climate change, and how we could address this. Since the studies were carried out by different researchers, looking at different populations, they have produced segments that don't exactly match between the countries. Nevertheless, there are enough similarities for there to be plenty of useful lessons. Of the three studies, the US and Australian are the more relevant, as they focus specifically on climate change. The British study looks at wider environmental attitudes and so is discussed more briefly.

Before getting into the detail of each segment, let's first look at the distribution of base, swings and critics.

The Australian segmentation,[31] carried out by academics from the University of New England and Griffith University, divides the population into five segments, which it names *Alarmed* (26 per cent), *Concerned* (39 per cent), *Uncertain* (14 per cent), *Doubtful* (12 per cent) and *Dismissive* (9 per cent). The first of these can be described as a base segment and the last two can be treated as critics, with the *Dismissive* more vocal than the *Doubtful* in their opposition to cutting emissions. This leaves the *Concerned* and *Uncertain* segments as the swings.

The US research,[32] carried out by the Yale Project on Climate Change and George Mason University, is similar to the Australian one. It found six segments, of which three are swings. The segments are labelled *Alarmed* (17 per cent), *Concerned* (28 per cent), *Cautious* (27 per cent), *Disengaged* (7 per cent), *Doubtful* (11 per cent) and *Dismissive* (10 per cent). As in Australia, the first segment can be thought of as the base and the last two as the critics.

That leaves the UK segmentation,[33] which was commissioned by the Department for Environment, Food and Rural Affairs, and found seven segments: *Positive Greens* (18 per cent), *Waste Watchers* (12 per cent), *Concerned Consumers* (14 per cent), *Sideline Supporters* (14 per cent), *Cautious Participants* (14 per cent), *Stalled Starters* (10 per cent) and *Honestly Disengaged* (18 per cent). While these don't correspond directly with the US or Australian segments, there are parallels. The UK *Positive Greens* are similar to the *Alarmed* segments in the other countries and can be seen as the base. The *Stalled Starters* and *Honestly Disengaged* can also be seen as the critics, mirroring the *Doubtful* and *Dismissive* segments in Australia and the US. Like those segments, one of the UK groups is less firm in their doubts about climate change, while the other tends to be more convinced that it is a hoax.

Separate UK research, which was based on underlying values rather than attitudes and environmental behaviors, found the population was roughly evenly split between those who prioritize ethical issues: those whose priority is how others see them; and those

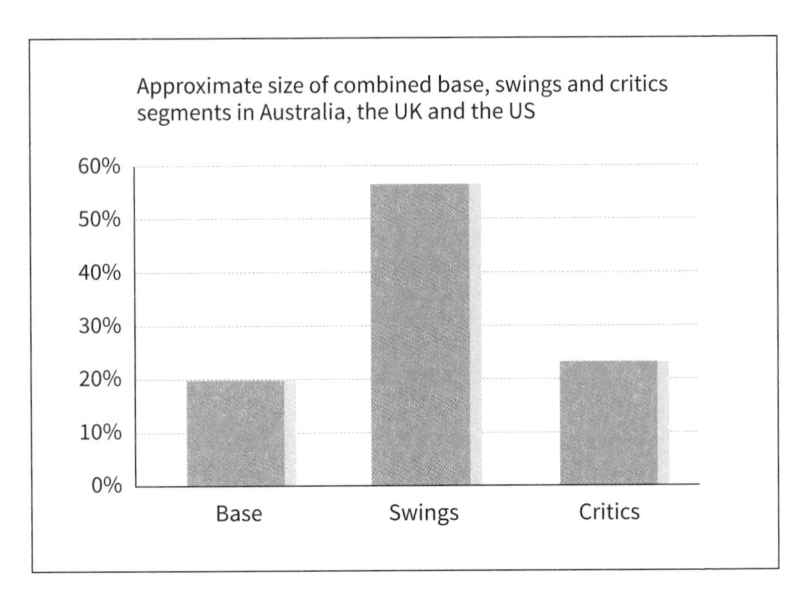

whose priority is safety and security.[34] Although that segmentation process was not directly comparable to the others, and it suggests there are slightly fewer swings, the results of that values-based approach are fairly similar to those of the other studies.

In each country we have a base segment, of between 17 per cent and 26 per cent of the population. These are the people who are most worried about climate change and strongly support action to address it. You may notice that this is much less than the proportion that, as we saw earlier, says climate change is real and caused by humans. That is because the people in the base segment are those who not only accept climate science and its implications but also tend to be vocal in supporting action to address the problem. Many of them have already made some changes to their own lives. A typical sentiment of people in the base segment was expressed by a participant in the UK study: 'We need to do some things differently to tackle climate change. I do what I can and I feel bad about the rest.'[35]

At the other end of the scale, the two critic segments make up between 21 per cent and 28 per cent of the population. This is more than the proportion who say that climate change is not happening or is not a threat. Indeed, only about half of these people – those in the most critical segment – consistently say they think the whole thing is a hoax. A quote from research in Australia exemplifies that segment's views: 'I have not heard anything from people that I know and trust and with the knowledge that leads me to believe that climate change will happen because of us. There appears to me to be people that are and will be making a lot of money out of this.'[36]

People in the other critic segment, the *Doubtful*, while tending to be unconvinced by climate science, are much less steadfast in their doubts and often say they could change their minds. Another participant in Australia summed up their view: 'I have no real way of knowing just who I should believe... So I can choose to blindly believe what somebody wants me to believe or I can wait until clear undisputable scientific facts are presented and then I can make an informed opinion.'[37] While they don't seem worried enough about climate change to count as swings, they aren't such consistent opponents of measures to cut emissions as those in the other critic segment are.

For the reasons outlined in the previous chapter, the two critic segments shouldn't be our priority when we are trying to improve the world's chances of avoiding disastrous climate change. So after that brief introduction we won't see much of them in the following chapters. But one point about those two segments is worth mentioning. As we saw earlier, there is a debate about whether those who say climate change isn't a problem should be labelled 'skeptics' or 'deniers'. Now we have seen how the critics can be split into two segments, perhaps that question can be resolved – one group appears skeptical about climate change but potentially open to persuasion, while the other seems determined to deny its significance, in the face of all evidence.

Let's return to the swings, who should be our priority. The swing segment that is more worried about climate change looks quite like a less committed version of the base segment. People in the *Concerned* segment generally accept that climate change is happening, caused by humans and a problem, but while they are worried about the threat, it is not often on their minds. Their knowledge about climate change is typically little better than average and they don't identify themselves as environmentalists any more than average. Compared with people in the base segment, they are not passionate about the problem and they generally haven't changed much about their lives to respond to it.

In general, the *Concerned* are slightly more likely than average to be female, younger, wealthier and urban. That said, it is best not to depend on demographics when thinking about any of the segments – it wasn't the basis of the research, and most of the segments are demographically similar to the general public.

While researching for this book I discussed climate change with people who belong to the two swing segments.[38] Talking about the issues with these swings helped me to understand better how people in the two segments view climate change. One person from the *Concerned* segment told me: 'It's not that I don't think it's important, or that I doubt the science, or that there are lots of people who care deeply about it, I just don't think it's likely to have that much impact upon our lives within our generation.' Another compared it with other issues: 'It matters, but in all honesty it feels like a more distant problem than, for example, cuts to public services and the rise of anti-refugee rhetoric. I feel those matter more.'

People in the second swing segment, the *Cautious*,[39] also tend to agree that climate change is happening and is likely to be a problem, but they are not particularly worried about it. They typically have not spent much time thinking about the issue and don't regard themselves as very well informed about it. The authors of the Australian study suggest the members hold contradictory opinions – they accept climate science but that doesn't match their views about the likely consequences. Demographically, members of this group are a little more likely than average to be male and younger.

One person from this segment told me: 'I can't say I regularly give it much thought', while another said: 'I think there are more pressing issues to be concerned about.' Some expressed doubt about the effectiveness of policies that seek to reduce emissions: 'I would be very wary about anything that involves taxes or money and skeptical that it does anything but raise tax or inadvertently penalize poor people. Faced with the intangible of climate change and tangible of day-to-day issues that I care about, I will always side with the day-to-day.'

A finding from the UK research can help us understand the *Cautious* further. It identified two separate segments that, between them, roughly correspond to the *Cautious* – but with an interesting difference between the two UK segments. People in one group tend to consider themselves to be knowledgeable about climate change and to have an ecological outlook, yet their concern is principally for the UK countryside rather than for the global climate. In contrast, members of the other UK segment are more worried about climate change but have fairly low knowledge about it and are the most likely to say that they would be embarrassed to be considered green. It may be that these contrasting views could both be found within the *Cautious* segment in the other countries.

The US segmentation also finds a small third swing segment – seven per cent of the public – that has thought very little about climate change and doesn't have strong views about it. This segment, the *Disengaged*, doesn't quite have a parallel in the other studies (rather than reflecting a difference about the US public, this may be the result of a decision by the researchers to create a distinct segment for the people who said they didn't have many opinions – about anything – rather than fitting them into other segments; this is a recurring problem in segmentation studies). They are similar to the *Cautious*, but spend even less time thinking about climate change and consider themselves to have less knowledge about it. As such, they say they don't know whether it is likely to be a threat to them personally. This lack of interest in climate change makes them harder to interest in the debate, and so it is more useful to think of the *Concerned* and *Cautious* as the principal swings.

While there are differences between the two swing groups, they are united by an important characteristic. Most members of both segments understand that climate change is caused by humans and will be a problem unless it is addressed. Yet they have little interest in the issue and often say they would oppose measures that would cut emissions. This isn't climate denial, since they don't dispute the reality of the problem. Instead, it's something equally challenging – climate apathy. Distinguishing between apathy and denial is vital if we are to target our efforts where they can make the most difference to opinion about climate change and measures that could reduce it. But so far, that distinction has usually been overlooked.

Addressing climate apathy among the swings, who make up about half the population, should be a priority if we are to increase support for the actions needed for the world to avoid disastrous global warming. The two swing groups seem persuadable to the conclusion that climate change deserves serious attention – they are certainly more winnable than those in the critic segments. And one lesson we can immediately draw is that, since those who are apathetic mostly already accept that climate change is happening and caused by human activities, tackling apathy doesn't depend on demonstrating the scientific consensus.

Instead, climate apathy is based on other factors that need to be addressed if the swings are to become, in their views of climate change, more like the people in the base segment. The coming chapters investigate the causes of apathy, including how it is inadvertently strengthened by the ways climate change is discussed, even by people who are trying to encourage action to address it. And we will see how we could persuade the swing segments that climate change matters to them and why it is worth supporting action to reduce it.

Already, from the segmentation research, we can begin to see the outlines of some of these factors behind apathy. Many of the people in the swing segments are aware that climate change is real and a problem, but they still haven't become particularly interested in it. To them, it has never been obvious why they should pay more attention to the problem. They have heard about climate change, have no reason to reject scientists' findings, and often have a feeling it is important – yet they generally recognize that they don't know much about the issue and don't spend much time thinking about it.

Some among the swings don't fully accept that climate change is a major problem. This is most common among those in the second segment, the *Cautious*. For some people, this lower concern is based on being more worried about other, local, environmental issues. For others, the attitude may be more emotional than rational – they accept the findings of climate science but just haven't allowed themselves to absorb its implications. Another challenge is the unwillingness of some swings to identify as green. Again, these people might have no objection to the evidence that climate change is caused by humans and will be a problem – they just find it difficult to change their lifestyle.

But before getting onto how we can challenge climate apathy, we first need to answer the 'why' questions. Why does public opinion about climate change matter? And why should we care about limiting global warming? The next two chapters address those questions.

3

Maps and roadblocks

Despite the 2015 Paris Agreement, the world is heading towards dangerous warming. Many countries are now cutting their greenhouse-gas emissions. But these measures won't be enough and some measures that will be needed to avoid disaster would be opposed by most people.

'If everyone does a little, we'll achieve only a little.'
– Sir David MacKay, Regius Professor of Engineering, Cambridge University[1]

In December 2006, the European Parliament agreed to major new regulations that put significant restrictions on businesses, with the goal of protecting the environment and improving the health of the 500 million people living in the European Union (EU). One estimate suggested the rules would cost the private sector up to 13 billion ($13.7 billion).[2] The regulations are known as REACH[3] and govern the manufacture and import of chemicals, covering a vast range of substances and placing a burden on companies to prove the chemicals they are using are safe. REACH was described as the most important EU legislation for 20 years,[4] but a survey found that a large proportion of manufacturers were unaware of the rules – and it is hard to imagine that many members of the public had heard of them either.[5]

The introduction of REACH shows that ambitious laws to protect the environment can be introduced without widespread pressure from the public or significant efforts by governments to win popular support.[6] It proves that governments can add regulation that increases the cost of everyday products without being driven to do so by a public clamor for action.

If that is the case, why should we care what the public think about climate change? Perhaps all these discussions about public opinion – how many people think climate change is real, how worried they are about it, whether they would back actions to cut emissions – are missing the point. It might be that governments can prevent disastrous warming regardless of whether the public is interested. Perhaps, as long as those in power see the need to address climate change and most of the public don't actively oppose action, governments can get on with what is needed. And, since majorities in major developed countries do agree that the problem should be addressed, governments might already have all the public support they need.

This chapter investigates that question. It looks at what the world needs to achieve if it is to avoid disaster, how successful it has been so far, what more will be needed, and whether a majority would support such measures.

The finishing line

To begin, we need to be clear about what the world should aim for. The goal of any action to address climate change comes down to limiting one of two things: the extent to which the planet heats up, or the consequences of the changes caused by that increased heat. Some measures are designed to achieve both, particularly in poorer countries, where programs to promote economic development are often designed to help people become better able to cope with a changing climate in ways that also avoid increasing emissions.

There is no serious doubt that the world needs to reduce its greenhouse-gas emissions, although there is no consensus about the precise level to which they should be limited. The 2015 Paris climate conference agreed to a goal of tightly restricting global warming.[7] The temperature increase that the Paris Agreement aims to avoid exceeding is the most ambitious limit that international governments seriously consider and is generally described as 'safe' – even though it would still bring some dangerous changes, especially to people in the world's poorest countries and low-lying islands. Yet some argue that this target is unachievable and that the world should aim for a more realistic, higher, level of warming. The disagreement isn't about the ideal maximum level of warming but is about the feasibility and costs of achieving particular targets.

Public support is part of that question of feasibility.

There is even less consensus about how far people and the natural environment should be helped to cope with changes to the climate that cannot be avoided (known as adaptation). That isn't surprising, as it is hard to quantify. The ways in which people are affected by climate change depend on many factors, including not only how their local environment responds to global warming, but also their vulnerability and resilience and how other people respond. For example, in parts of east Africa, changes to rainfall might allow crops to grow on land that was previously used only by nomadic herders;[8] if this happens and settled farmers move in, what do the nomads do? Some attempts are being made to develop global adaptation goals,[9] although these will never be as easily measurable as an overall level of warming. For the purposes of thinking about what adaptation efforts are trying to achieve, we should stick with a simple definition that they seek to stop people's quality of life deteriorating as a result of the changing climate.

With this in mind, let's look at how successful the world has been in addressing climate change. If everything is going well, perhaps climate apathy isn't blocking progress on limiting warming or helping people adapt to changes.

International commitments to limit the emissions that cause global warming effectively began with the agreement to create the UN Framework Convention on Climate Change (UNFCCC) at the Earth Summit in Rio de Janeiro in 1992. In Rio, governments agreed in principle to limit their emissions, but it wasn't until a conference in Kyoto five years later that any of them accepted specific targets for emission cuts. The resulting Kyoto Protocol runs until 2020, when it is due to be succeeded by the Paris Agreement.

These attempts to reduce emissions can seem like failures. Greenhouse-gas emissions have been increasing for years, so emissions over the last decade were greater than over any previous decade in human history. This has been a consequence of economic growth – in general, the more economic activity there is, the more emissions are produced. If further growth is to be reconciled with avoiding dangerous climate change, economies must become cleaner, so that each unit of economic activity produces lower emissions – the 'emission intensity' must decline. If total emissions are to fall, emission intensity has to decline more quickly than economic activity increases.

But perhaps this pessimism blinds us to where the world is making progress. For a start, we don't know how much worse emissions would have been if there hadn't been any efforts to restrain them. And, while the world's annual emissions have generally been rising, in each year from 2014 to 2016 global emissions appear to have barely increased at all.[10] This suggests that emission intensity has been falling at about the same rate as the global economy has been growing.

Nevertheless – and progress of recent years notwithstanding – it is hard to avoid the conclusion that efforts to reduce emissions have been slower than needed. If the world is to limit warming to relatively safe levels, it will need to do much more than just stop increasing emissions. The near-absence of emissions growth since 2014 is encouraging, but it is only a start. If I was in a car hurtling towards the edge of a cliff, I would be glad if the driver stopped accelerating, but I would still be desperate for them to step on the brake.

We see a similar picture when we look at progress on adapting to climate change. The world's poorest people will generally suffer the most from a warming world, as they are least able to cope with extremes and tend to live in the places that will face the fiercest heat waves, droughts, floods and other disasters.[11] The governments of many developing countries are introducing measures aimed at responding to the changing climate and preparing for further changes to come. Wealthier governments have said they will distribute at least $100 billion a year for climate-change programs from 2020, around half of which is intended to go towards adapting to changes. Many governments of richer countries also claim to be addressing the threats to their own people from climate change, with measures like flood defenses and changes to building rules to make homes better able to cope with extreme heat. Yet, despite these efforts, increasingly extreme weather is likely to cause far more damage than adaptation plans seem ready to address. The cost of adapting to climate change could be $500 billion a year by 2050[12] – around 10 times the amount that richer countries say they will offer to help developing countries.

So, even if climate policies haven't been a complete failure so far, future measures will need to go much further. This is where public opinion may be an obstacle. We can't assume that what has been achieved is evidence that the public will support the measures that will be needed in future. It might be that the world has done only the easy things and that what is to come is beyond what the public would accept, unless attitudes change.

To understand this, the rest of this chapter explores where measures to limit and deal with climate change are needed, and which of these might come into conflict with public opinion. No-one knows precisely what it will take to limit warming or how the world can best protect against the worst effects of warming. Rather than pass judgement on what exactly will be required, the chapter outlines some of the most significant measures that are seriously proposed. They might not all be necessary, but many of them will be.

How to avoid disaster

To have a good chance of keeping warming to a reasonably safe level, the world needs to reduce its greenhouse-gas emissions to zero well before the end of this century. In fact, overall emissions will almost certainly have to be negative – more greenhouse gases will have to be

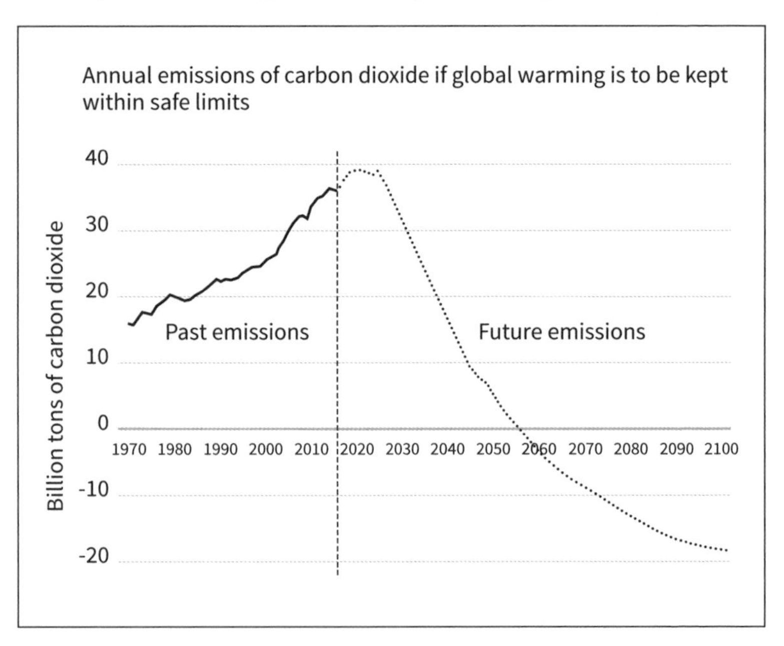

absorbed each year than released.[13] This is a phenomenal challenge. The chart[14] above, which shows emissions over the past 40 years, and the maximum emissions possible over the rest of the century to keep warming within safe limits, reveals how sharply recent trends would have to be reversed. The progress of the past few years in stopping the annual increase in emissions is trivial in comparison.

Where will these cuts need to come from? At the moment, around 29 per cent of emissions in richer industrialized countries[15] comes from activities that are invisible to most people, such as iron, steel and cement production, and the extraction and processing of energy stores like coal, oil and gas. One notional way of cutting these emissions would be through regulations or incentives to reduce the emissions intensity of the processes, so the industries continue functioning but with less impact on the climate. Another option would be to leave the processes unchanged but to reduce public demand for goods and services that depend on them – perhaps by greatly increasing recycling and the reuse of materials.

The next largest source of emissions is transport (27 per cent of the total), of which over 80 per cent is from roads.[16] A simple answer here would be to move around less, but other measures could also help. While it might seem that vehicles are rapidly becoming more efficient, in fact there has been almost no reduction in greenhouse-gas emissions, per mile, since 2010.[17] This suggests that a switch to zero-emissions vehicles, like electric cars, offers the best chance of cutting emissions from road use. Increasing public transport at the expense of car use might also cut emissions, although the difference would be smaller if cars were cleaner. Flying presents a greater challenge, as discussed later in this chapter.

About 17 per cent of emissions come from residential buildings, with a similar proportion again from the commercial and public sector. Of a building's lifetime emissions, around 15 per cent are from its construction,[18] so could be addressed through regulations or incentives that apply before anyone moves in. But most emissions come from the energy consumed during a building's use. Heaters and air conditioners might produce lower emissions if they are electric and the electricity is generated by clean sources. Efficiency would also help. Emissions from heating or cooling can be reduced with better insulation and more efficient devices.

Indeed, pretty much every electricity-using appliance can be made more efficient, from light bulbs to fridges and vacuum cleaners. There is some evidence that people respond to reductions in the cost of running an appliance by using it more, or by spending the money on other high-emitting activities; this is known as the rebound effect. But this rebound generally does not appear to be greater than the savings. Rebounds may reduce benefits, but they do not seem likely to reverse them,[19] so improving efficiency seems worthwhile from an environmental perspective.

The other major emission-producing sector is agriculture (10 per cent of the total), mostly linked with raising livestock. These emissions could partly be reduced with changes to farming practices, such as in the treatment of manure. The emissions would also be cut if people ate less of those things whose production is a major source of greenhouse gases. This would mean reducing demand for animal products, particularly red meat and dairy, as producing them causes especially high volumes of greenhouse gases, including from forest clearance and the belches of cows and sheep (yes, really[20]).

Several of these potential emissions cuts depend on a cleaner electricity supply. Electricity generation alone currently represents around 31 per cent of emissions in richer countries (this is included in the sources of emissions already listed). A major problem at the moment is that many cleaner sources only function intermittently, such as when the sun is shining or the wind is blowing. This could partly be addressed with more flexible or large-scale grids, such as one covering the European Union, that allow electricity to be shared across areas with different weather patterns. Another option might be the import of electricity from solar power that is generated in places where sunshine is reliable, like North Africa, although the practical challenges of doing this have so far been insurmountable[21] and exports of other natural resources from countries in the Global South to those in the North (such as oil and diamonds) have generally brought little good – and much harm – to most people in the exporting countries.[22] Nuclear power represents a relatively low-carbon option, but in its current form it is expensive, can't be rapidly increased to meet peaks in demand, and faces problems of waste and safety. A more robust answer to the problem of intermittency would be the creation of better and cheaper ways of storing power, so energy from low-carbon sources can be stored at times it is generated but not needed. For now, the only affordable solution to the problem of intermittency is the burning of fossil fuels, which can more easily be switched on at short notice. The widespread replacement of coal with natural gas in recent years in richer countries has cut emissions, but gas is still too dirty in the long term to avoid dangerous warming. The only long-term future for natural gas as a major power source would be if it is combined with technology to capture and store emissions, which is still expensive and unproven on the scale that would be needed.[23]

Another way the world could limit global warming would be deliberately to change the planet in ways that counteract emissions, known as geoengineering. These measures come in two categories: those that extract greenhouse gases from the atmosphere, and those that reduce the amount of energy from the sun that the planet absorbs.

Extracting greenhouse gases from the atmosphere seems less controversial than the other form of geo-engineering as it could reduce climate change without requiring the world to stop doing the things that produce emissions. Some of the ways of doing this are based on increasing the amount of photosynthesizing plants, so more carbon dioxide is absorbed and stored in the plant. This includes spreading nutrients in the sea to fertilize phytoplankton growth, and allowing the dead plants to sink to the ocean depths; planting many more trees; or creating a mix of charcoal and soil that can store carbon dioxide for centuries. Generating electricity by burning energy crops and capturing the greenhouse gases would have the same effect,[24] as the process would extract carbon dioxide from the atmosphere – this approach, known as bio-energy with carbon capture and storage, or BECCS, is a frontrunner in this area and is sometimes used in modelling of future emissions. Other measures draw on chemical processes, including breaking down rocks to increase the rate at which they absorb carbon dioxide; and using artificial methods to extract carbon from the air.[25] Most modelling of how the world can avoid extreme warming concludes that there will have to be extensive use of these technologies so that annual overall emissions can fall below zero.

But while carbon-dioxide removal might appear to offer a get-out-of-jail card, allowing the world to stop worrying about emissions, this is unlikely to be the case, for two reasons. First, applying them widely would be far from easy. Most of the techniques would be hugely expensive. Some, including BECCS, would require a vast proportion of the planet's surface and so would compete for land with agriculture and wildlife, pushing up food prices and threatening rapid and widespread extinction. Others, such as fertilizing plankton growth, could cause unpredictable harm to life in the oceans.

Second, even if the cost and unwanted side-effects of these technologies could be reduced, they would be needed in combination with emissions cuts, not instead of them. No-one yet knows exactly how much carbon dioxide these approaches could absorb, but an upper estimate used by climate models suggests it could reach around 20 billion tons of carbon dioxide a year by the end of the century.[26] Projections of future emissions that keep warming to safe levels suggest that overall emissions would need to be well below zero by 2100, anyway – the world should be absorbing more carbon dioxide more than it is producing.[27] Assuming the world succeeds in removing as much carbon dioxide from the atmosphere as currently seems feasible, humans could produce a maximum of around four billion tons of carbon dioxide a year by 2100.[28] For comparison, human activities emitted around 36 billion tons of carbon dioxide in 2016,[29] and if the world does little to cut them, annual emissions could be well over 100 billion tons a year by the end of the century.[30] So these technologies might offer a little breathing room in the race to cut emissions, but not much more than that.

The second set of geoengineering measures is both easier and even less attractive. Options in this group include increasing the reflectivity of clouds and putting mirrors or dust in space to block sunlight. Perhaps the most plausible would be the injection of particles into the stratosphere, mimicking the effect of major volcanic eruptions. This could reduce global temperatures enough to counteract global warming entirely, but it may well bring with it enormous problems.[31] The stratospheric 'veil' would, researchers suggest, change global weather systems, probably in ways that would make some people's lives harder, such as by disrupting monsoons in Asia. This would mean that many people would suffer from the consequences of a choice made in other countries, which would create a powerful grievance that could destabilize international relations. While climate change is a product of human activities, no-one set out to create it in the way they might seek to alter the planet through blocking the sun's energy. A further problem is that reducing sunlight does nothing to stop carbon dioxide concentrations making oceans more acidic. The fact that this form of geoengineering could be done cheaply (a stratospheric veil could be created for as little as a few billion dollars a year) and with existing technology means it will offer a way of reducing warming if the world fails to cut emissions. But it seems best to consider it only as a desperate last resort, if warming would otherwise be so catastrophic that it would be better to face the consequences of knowingly disrupting global weather systems.

Measures to help people adapt to climate change can be split into two categories, which could be called 'enabling' and 'restrictive'. Enabling adaptation helps people to carry on living as they were before the climate changed, with measures to respond to floods, heat waves and droughts. The main difficulty with enabling adaptation is that it is hard to do enough of it. That would be a greater problem in poorer countries, but even in richer ones the impacts of extreme global warming would be so great that it wouldn't always be possible for people to carry on as if it wasn't happening.

This is where restrictive adaptation comes in. Since many people wouldn't always be able to carry on living in the same ways in the face of extreme global warming, sometimes adaptation may need to limit what they can do. Where flooding is likely to be a problem, the public bill for repairing flood-damaged infrastructure could be large. After floods in 2014, the UK prime minister declared that money was 'no object' for the repairs.[32] The expectation of similar largesse may discourage people and businesses from paying to prevent flooding in the first place. But as climate change worsens, governments won't be able to afford to be so generous and they might be forced to tell people in areas prone to disasters worsened by climate change that they won't get help unless they move.

Governments might also discourage people from making decisions that will lead to them needing public support in the future. For example, rules for new buildings could ensure homes don't overheat during hot summers, or they might stop new homes and businesses being built in places that will be more at risk once the planet warms.

Having seen the kind of actions that are likely to be needed to avoid extreme warming and to deal with its effects, the remainder of the chapter considers how far public opinion may limit what can be achieved.

Why public opinion matters

Many measures to cut emissions could in theory be introduced with little or no attention to public opinion. Such changes to industrial processes, agriculture and building regulations might increase prices for consumers, but the cause of those increases perhaps wouldn't be obvious. It isn't an attractive model of democratic government but, as we saw with the EU's chemical controls, politicians can introduce some major regulations without engaging the public.

But even if we were comfortable with governments addressing climate change without much public engagement, back-room measures like these still wouldn't be enough. There is no realistic path to avoiding dangerous global warming that doesn't involve most people in richer countries changing aspects of their lives. In an age where elections and referendums are going against those that are seen as distant elites, such an approach feels even less sustainable.

So, how much of an obstacle is climate apathy? Across the range of measures that might be needed, we can distinguish three different sets of actions, placed along a scale from 'many winners' to 'many losers'. The first is fairly easy to introduce; the second is harder; the third is harder still.

1. Wide support

The first set of measures doesn't require the public to make major changes or to accept significant costs. Many of them are based on the use of new technologies and have the potential not just to offer lower emissions but also to bring other benefits. Some have already been legally mandated in many countries, such as rules on the power consumption of light bulbs and other appliances. Others are becoming more popular without being compulsory, sometimes supported by subsidies, notably for electric cars. Still others may be needed but aren't yet ready for widespread use, particularly technologies that can cut emissions from home-heating systems, such as efficient electric boilers and heat pumps.

What makes these actions achievable in terms of public opinion is that they can bring benefits unrelated to climate change, without adding significant costs. When new technologies are better than the ones they replace, such as being cheaper to run or more effective, resistance to them tends to be limited. The EU's ban on incandescent light bulbs was a good example of this. The losers are the small number of people with a profound attachment to old-style inefficient bulbs and those who are poor enough not to be able to afford more expensive light bulbs even though they are cheaper to run. But most people were winners as the change meant manufacturers focused on new bulbs, which were cheaper to run. There was no need to refer to emissions reductions to justify the change, and, without much fuss, it seems to have reduced energy use.[33] Nevertheless, the fact that more efficient bulbs were legally mandated indicates that even these technological advances don't always spread by market forces alone and governments still feel they have to promote them. Individuals don't make decisions on a purely economic basis and whether or not they can be seen to win or lose from a change depends on more than just a financial calculation.

Only the most hostile of the segments tend to resist measures in this group. Opposing changes that not only cut emissions but also make life cheaper relies on a determination to resist anything linked with climate change. As such, opposition is mostly limited to the two segments of critics. In fact, it might come only from the 9 to 10 per cent in the most critical segment, as their hostility to climate policy conditions them to look out for measures such as these. So, even if public concerns about climate change were to remain unchanged, it shouldn't be difficult for governments to introduce many measures in this first group. Nevertheless, after its success with light bulbs, the

EU backed out of mandating similar standards for toasters, apparently because of fear of media criticism,[34] suggesting that this limited opposition can sometimes still be enough to stop action. Despite its smallness, the opposition can influence government decisions in this way because of the effectiveness of groups that deny the threat of climate change. This reflects the importance of demonstrating how rare such views are and challenging the credibility of people leading the opposition when they claim to represent either scientific evidence or public opinion.

2. Some support, some opposition

The second set of actions is harder to win support for. Unlike the first set of measures, there are visible costs associated with them. What's more, it is harder to identify winners in the absence of public recognition that avoiding dangerous climate change is a benefit.

The most important among the actions in this group is reducing emissions from electricity generation. This is essential, not only because electricity generation currently produces around a third of emissions in richer countries, but also because many of the other measures for cutting emissions, such as a switch to electric vehicles, depend on a clean power supply. In fact, much of the work of cutting emissions can be boiled down to two steps: electrify as much as possible and make electricity clean.[35] But it is not cheap to build new power plants or electric versions of products that currently use fossil fuels, particularly where existing fossil fuel-based infrastructure is still viable. While renewables are, in some places, now the cheapest source of power, they are generally still more expensive than their dirtier rivals.

Another measure in this category is reducing energy consumption of homes, through better insulation in colder countries and less need for air conditioning in hotter places. This would require improvements being made to existing homes as well as changes to building regulations. Enabling adaptation also belongs in this category, with public funding ensuring people can carry on living as they were already, for example with improved flood defenses.

Like the first set of measures, these actions tend to bring benefits to the public which aren't linked with climate change. Closing coal-fired power stations improves air quality, potentially saving hundreds of thousands of lives each year;[36] switching to solar, wind and water power reduces reliance on imports and creates jobs; reducing the need for home heating or cooling cuts domestic energy bills.

But such measures also involve costs and face public challenges. So long as clean power is more expensive than coal and gas, a switch to it has to be paid for through taxes or bills. People living near proposed power-generation facilities sometimes strongly oppose their construction. The development of onshore wind farms exemplifies this – while polls repeatedly show their popularity,[37] strong opposition from a small number of people, backed by some newspapers, can overwhelm widespread but shallow support. Similarly, while it might be possible to build homes that require little-to-no heating or cooling at minimal extra cost,[38] there is certainly a cost to improving existing homes, which might have to be met by people who don't directly benefit. In that case there are many losers: people whose bills or taxes increase to pay for others' home improvements, or who experience cuts in other public spending. While there are also winners – people whose homes are upgraded – it may be that the costs matter more to the losers than the gains do to the winners. In the absence of widespread agreement that emissions need to be cut, it might be difficult to generate sufficient enthusiasm to overcome this opposition.

Since support for this set of actions seems to depend on concern about climate change, we can anticipate that fewer of the segments would currently back them than would support the first set. The critics, roughly 25 per cent, would be unlikely to support them, while we might expect the base, about 20 per cent, to be in favor. How about the swings? The *Concerned* seem likely to be favorable – they are worried about climate change and want action to address it, but haven't been motivated to take action themselves. Since these measures don't require them to do much except pay a bit more, their support may be forthcoming, as long as the cost isn't too high. The other half of the swings, the *Cautious*, are more of a challenge. They can be suspicious of policies that increase costs in order to cut emissions. So all in all, we might predict that slightly less than half of the public would support these measures. That estimate is backed up by polls about proposed emission-cutting measures – around 40 to 50 per cent typically say they are prepared to pay more for lower-carbon energy sources.[39] This seems a weak basis for such vital measures, particularly when, as with opposition to wind farms, opponents often appear to care much more than all but the most enthusiastic supporters do.

3. More opposition

The challenge looks even more daunting when we turn to the third set of actions. For these measures, the burden to act – or forego action – is on the public. While the government can promote or mandate them, their success often depends on many people making changes that go beyond paying more. And the measures don't seem to offer compensatory benefits beyond avoiding dangerous climate change, making it easy to identify losers and harder still to see winners.

The measures in this category typically involve restrictions and there are two particularly difficult challenges. The first is flying. The UN's aviation agency announced a plan in 2016 to keep global aircraft emissions constant at their 2020 level (in contrast, nearly every other sector plans to dramatically cut their emissions). To achieve this when the industry expects the number of flights to increase rapidly, it is counting on various measures that, it says, will reduce the emissions from each flight, such as planes becoming more efficient and jet fuel being mixed with cleaner biofuels.

But this plan doesn't solve aviation's emissions problem. One issue is that it is optimistic, to the point of implausibility, about how far the industry can cut emissions per flight. Among the elements that stretch belief is the suggestion that the world will build 170 biofuel refineries every year for 30 years, at a cost of up to $1,800 billion.[40] The biofuel that it depends on would also have to compete with the land and fuel that may be needed to absorb carbon dioxide as well as with the land needed for food and by wildlife. But even if, against the odds, the industry succeeded with every element of its optimistic plan, aviation emissions would still use up around 12 per cent of the total that the world could produce before 2050 if it is to avoid dangerous warming, compared with around 2 per cent at the moment (if the plan fails, flying would take up around 27 per cent of world emissions by 2050).[41] Other sources of emissions – electricity production, industry, buildings and so on – would have to shrink further to create room for flying. And this problem becomes even greater in the second half of the century. By the end of the century, humans might be able to emit a maximum of four billion tons of carbon dioxide a year, as outlined above; aviation's most optimistic plan suggests it would use around 18 per cent of this unless it can find a way of cutting its emissions even further.[42] The rest of the world would have to work ever harder to balance out the carbon dioxide produced by each flight, unless passenger numbers began to fall.

Another of the particularly hard challenges is meat and dairy consumption. Food production is responsible for a large share of global emissions – perhaps a quarter, depending on what is counted – and most of this is from livestock. Models that limit warming to safe levels suggest the gases produced by livestock don't have to fall quite as much as those from burning fossil fuels – between around 33 and 44 per cent by the end of the century compared with their 2010 levels.[43] But the world is still far off course in achieving even that. Growing demand for meat and dairy would increase greenhouse gases – a 2016 study suggested that emissions from food production will rise by 51 per cent by 2050 unless diets change.[44] In contrast, a worldwide switch to a vegetarian diet would cut food-based emissions by up to 55 per cent while a vegan diet would cut them by up to 70 per cent. A switch to lower-meat diets that are not fully vegetarian or vegan would also cut emissions, though not by as much. Most of these savings would come from the reduction in red-meat consumption. This is a global problem – as people move out of poverty they tend to eat more meat – but it is hard to see how overall emissions can fall below zero if people in richer countries don't change their diets.

In both food production and flying, it is possible that some technology will come to the rescue; for example, a clean but powerful biological jet fuel that can be grown in massive quantities, or affordable and appealing artificial meat. But no such fix seems near. And, without that kind of solution, it is hard to escape the conclusion that the continued expansion of flying and meat consumption will make it much harder for the world to avoid dangerous warming. There is a similar challenge with shipping, which currently represents 2.5 per cent of global emissions: without measures to change course, those emissions are expected to increase by up to 250 per cent by 2050.[45]

Winning support for these measures will be difficult. As with the second set of measures, the strongest justification for them is that they would reduce climate change. Yet, unlike those measures, these don't obviously bring secondary benefits such as improved air quality, new jobs, or homes that cost less to run. Some people may consider eating less meat to be no sacrifice, but this is clearly not true of most people, for whom meat consumption is an important part of an enjoyable life. All that most people would see with this category is restrictions and sacrifices. If someone isn't convinced of the need to address climate change, it is not obvious why they would accept these sacrifices.

As public opinion currently stands, it is hard to see support for these measures stretching beyond the 20 per cent in the base segment, the *Alarmed*. Members of the most supportive of the swing segments, the *Concerned*, tend to be unwilling to make significant sacrifices to address climate change. In fact, not even everyone in the *Alarmed* base segment seems supportive. There is no sign of meat consumption falling in richer countries – although there may be a switch away from red meat, which would reduce emissions[46] – while the global growth in aviation shows that even if some people are voluntarily holding back from flying it is not nearly enough to balance the increasing enthusiasm of others for it. Perhaps those most willing to act are quietly waiting for collective restrictions on high-emitting activities but are reluctant to make unilateral sacrifices. Nevertheless, there is clearly a long way to go before more than just a small minority would agree to such restrictions. If technological advances don't save the world from needing these restrictions, public resistance will be an enormous barrier – but one that will have to be overcome if the world is to avoid dangerous warming.

This chapter began by questioning whether public opinion matters. To answer that, we should imagine what the world could achieve if public opinion remains unchanged. To avoid dangerous warming, the world will have to reduce emissions enough that measures to extract carbon dioxide can absorb more gas than human activities produce. Some of the measures to do this come at little cost and bring clear benefits unrelated to climate change. For these, public opinion doesn't seem likely to be an obstacle. Other measures would be more difficult as they impose costs on the public, even if these costs would be balanced by some benefits and don't require most people to change much about their lives. And a third set of measures presents the greatest challenge. Pressures to reduce flights and cut meat consumption would put a direct burden on the public and, excluding averted climate change, bring few benefits to balance those costs.

Overcoming climate apathy will be essential if governments are to introduce these emission-cutting measures. Even if governments set ambitious targets without engaging the public, meeting those targets will affect everyday life in noticeable ways and will only be possible with public support. But, as the views of the segments currently stand, winning this support requires a change in public attitudes. Over the rest of the book, we will see how opinion about climate change is formed and how those views could be influenced. Before that, though, the next chapter addresses the second part of why this debate matters – what will happen if the world fails to limit emissions.

4

The stakes

Even rich countries risk disaster from climate change. If the world radically cuts emissions, rich countries will face more extreme weather but would be able to cope. If it only meets its current emissions pledges, those extremes will be more destructive. And, if it fails to cut emissions, no country will escape catastrophe.

'A society's fate lies in its own hands and depends substantially on its own choices.'
– Jared Diamond, Professor of Geography, University of California, Los Angeles[1]

Why does this book have a chapter on the consequences of global warming? Most people with an interest in the subject could come up with a list of what is likely to happen as the planet heats – floods, heat waves, storms, rising sea levels, melting ice caps and extinct polar bears. But, beyond being able to say what, in general, we can expect from a hotter planet, most people would struggle to describe what different levels of global warming would mean in practice.[2] That problem is central to this book. As we will see over the coming chapters, one of the factors that limits concern about climate change is that many of the swings don't see how warming would affect them. And if many of those who are more worried about the problem can't say what is predicted to happen when the world warms, they are going to struggle to influence the swings.

So the aim of this chapter is to show how climate change will affect people living in some of the world's richest countries. This is not to suggest that those people will suffer the most from climate change. Poorer people, and particularly those in poorer countries, will be the main victims of global warming. This chapter focuses on the consequences of climate change for richer countries only because the purpose of the book is to find ways of influencing the opinions of people who live in those high-emitting countries, and local impacts are more likely to attract their attention.

The chapter looks at the likely effects of climate change this century, with three possible scenarios viewed from 2100, 85 years after the Paris Agreement was signed – the best case, where the world gets serious about cutting emissions and limits warming as much as possible; a middle ground where the world puts some effort into cutting emissions but fails to make the tougher sacrifices; and a scenario where the world does little, emissions don't fall and the planet rapidly heats up. Some readers may wonder what these scenarios correspond to in terms of average global degrees of warming. For reasons discussed in the next chapter, I am avoiding referring to average global temperature change. But for readers who want to put figures on the scenarios, they roughly correspond to temperature increases, by 2100, of 1.5°C/2.7°F, 3°C/5.4°F and 4.5°C/8.1°F compared with pre-industrial levels.

No projections of the consequences of climate change are certain, for two broad reasons. The first is that scientists can't say exactly what the consequences will be of a particular increase in global average temperatures. For any level of planetary warming, researchers can attempt to calculate how regional and local weather systems might respond, but these calculations are probabilistic. In addition, the consequences of changes in weather systems depend on other human activities. For example, the number of people whose homes are destroyed by wildfires is influenced by population growth and decisions about planning and land management.

The second major uncertainty comes from the relationship between emissions and warming. A fundamental question in climate science is how much the planet will warm by as a result of a given increase in greenhouse gases. Scientists' estimates of this 'climate sensitivity' fall within a range. This is further complicated by the potential for a certain amount of warming to trigger the release of more greenhouse gases – a vicious cycle that is discussed later in this chapter.

As far as these uncertainties allow, the descriptions of the consequences of global warming in this chapter are based on climate scientists' current understanding of what will happen as the planet warms. As more advanced models continue to be developed, these findings will be refined. But scientists have done enough work on these topics that it would be a great surprise if many of them were overturned.

A bit harder

12 December 2100: The 2015 Paris Agreement, adopted 85 years ago today, was a turning point in global efforts to avoid dangerous climate change. After years of isolationism, governments had realized that the only way to avoid disaster was for every country to cut its greenhouse-gas emissions. World leaders pledged not only specific cuts for the following 15 years, but to increase their commitments every five years, until emissions fell fast enough for global warming to be limited to little more than the temperature rises that had already been seen by the early 21st century. When the deal was signed, some said it was unachievable. But, with a combination of determination and technological advances, the world proved them wrong.

How would the world's climate change if it limited emissions so tightly?

The most visible sign of climate change to affect the UK by the early 21st century has been flooding. Even in this ambitious scenario, deluges would hit more of the country. In the second half of the century, the proportion of homes that would be exposed to flooding would increase by 40 per cent.[3] Many places that already flood after heavy rain would be likely to flood more often. Being flooded once is bad enough, and it can take years for individuals and communities to recover. If floods start coming around more often, so people have barely finished repairing the damage before they are hit again, many will question whether they can carry on living in the same place. Anyone who owns a home that becomes more flood-prone will face a difficult decision – stay, and endure repeated inundations, or sell, and accept the loss of value that comes with the rising waters.

This increased flooding would be a problem, but we shouldn't exaggerate the threat it poses. It would be a worsening of a current risk rather than a threat to the country's way of life. The people whose homes and businesses flood more often would need help and the insurance and planning systems would need to be updated. Even with the warming experienced so far, the UK's flood defenses often seem inadequate. But at the moment the UK government spends less than 0.1 per cent of its budget on flood defense[4] – even if this were to increase several-fold it wouldn't consume an unsustainable proportion of public spending.

The UK might see some benefits from a small amount of global warming, which would partly offset costs like flooding. This could include improved agricultural productivity. The increased temperatures may lengthen growing seasons and mean that more crops could thrive in the UK.[5] The UK's wine industry might already be benefiting from the changing climate, and conditions could improve further with a small temperature increase.[6] It is possible that increased carbon-dioxide concentrations would help plant growth as well, although the response would vary between plant species and depend on the other factors such as how much water is available.[7] The boost to agriculture might be limited by increased drought and the arrival to the UK of new pests, but on balance it seems likely that crop growers would see some benefit if emissions are limited this tightly. So all in all, a very small amount of further warming would bring problems for the UK, but these would be partly offset by separate benefits.

Chicago is better known for wind than for heat, but in 1995 it was struck by a heat wave that killed 739 people.[8] Temperatures were at least 32°C/89°F every day for a week and on two consecutive days reached 39°C/102°F. Heat like this was exceptionally rare in late 20th-century Chicago – more than two decades later, the city hasn't experienced another comparable heat wave. But towards the end of this century, even under this optimistic scenario, Chicago could expect to face that kind of heat every other year. Heat waves that are even worse would strike the city once a decade.[9]

Temperatures around that level might not sound particularly dramatic. Much of the world's population regularly experiences hotter temperatures now, without many ill effects. But the death toll from high temperatures isn't just a factor of how hot it is – what matters is how well prepared people are for the heat. The reason so many people died in Chicago's 1995 heat wave wasn't just because it was hot – it was also because the city couldn't cope. There wasn't an effective warning system, power supplies failed, healthcare wasn't up to the task, and many people couldn't afford air conditioning and were too afraid of crime to open their windows.

As the world warms, people will probably become better at coping with heat. When temperatures in Chicago increase, and this kind of heat wave becomes regular, the city will adapt. It will learn from cities that are more used to heat and will build public shelters for when it is too hot for the elderly and ill to stay at home, invest in warning systems and healthcare, and perhaps subsidize home improvements for the poor. This will take money that could have been spent on things like schools, but if the city manages the climate shift well it could avoid a significant death toll.

To many countries outside the tropics, such as Canada, the UK, the US and much of Australia, the direct effects of this scenario will be familiar from the weather extremes of recent years. That is not a surprise – the world has already warmed, by the mid-2010s, more than halfway to where it would ultimately settle if emissions are controlled this tightly. With good preparation, richer countries should be able to manage those changes so they don't suffer much more damage than they do already.

But people in richer countries won't only be affected by the changes that happen in their own areas. Because of the knock-on effects of changes that happen in other parts of the world – which will be hardest hit by this relatively small amount of warming – even richer countries will be more affected than might be expected.

The Syrian civil war may be the greatest humanitarian disaster of the century so far. In the first four-and-a-half years of the conflict, more than a quarter of a million people were killed and over 11 million forced from their homes, 4.5 million of them into other countries.[10] Most refugees only went as far as Turkey – at least at first – but in 2015 alone 363,000 sought asylum in the European Union.[11] The crisis led to the suspension of free movement within parts of the EU, eating away at one of its fundamental principles. The conflict also created the conditions for ISIS to seize control of large parts of Syria and Iraq, forcing millions of people to live under their brutal rule and giving the group a base to plan and encourage terror attacks around the world.

There is some evidence that climate change was a factor behind the disaster. In late 2006, Syria entered a prolonged drought that was the worst to hit the region since records began.[12] Between 2007 and 2008, prices of wheat, rice and animal feed more than doubled and agriculture in many parts of the country collapsed. As many as 1.5 million people left their homes to escape the drought – in some areas school enrolment fell by 80 per cent – and the urban population rocketed, growing by more than half between 2002 and 2010.[13] This migration created a growing threat of instability, with more people in crowded and illegal settlements on the fringes of cities that were barely able to cope with them. Then, just as the drought was coming to an end, the region was hit by further weather-related disasters.

These disasters weren't in Syria or even in the Middle East, but across Eastern Europe and East Asia. In Russia and Ukraine, drought, heat and fires cut wheat production by 33 per cent and 19 per cent respectively. In China, the world's largest wheat producer, drought in the wheat-growing east prompted the government to start importing wheat in anticipation of a shortage.[14] As a result of these weather extremes, and governments' responses to them, international wheat prices more than doubled from summer 2010. Many countries in the Middle East are wheat importers and were vulnerable to these price rises. While the Arab Spring was a protest against authoritarian rule, in places it was also explicitly a response to food prices – in Tunisia and Egypt protesters waved bread during demonstrations.[15] The protests that started in those countries then spread to Syria.

Putting this together, it is not a stretch to say that climate change was a factor behind the disaster in Syria and the resulting refugee crisis. There is evidence linking at least some of these weather events with climate change.[16] The outbreak of anti-government protests in the Middle East seems, in part, to have been triggered by the consequences of those events. Of course, other factors were also necessary and climate change alone wasn't sufficient. Without decades of poor governance, authoritarian rule and rapid population growth, the Arab Spring might never have happened. But without climate change it might not have happened, either.

The catastrophe to befall Syria reflects how even a small amount of global warming can reverberate around the world. Not only does climate change have direct victims, but, when a country is badly governed, has many people who don't have reliable incomes or savings, and is exposed to sudden changes in international prices, global warming can suddenly tip it into disaster. The Syrian refugee crisis and rise in international terrorism, which may themselves have been catalyzed by the effects of local and distant climate change, show that such disasters aren't confined to the countries that fall apart.

Too wet, too hot, too dry

12 December 2100: Viewed from 85 years later, the 2015 Paris Agreement was an important step forward in global efforts to avoid dangerous climate change – but there were still many backward steps to come. After years of isolationism, it seemed that governments had realized that the only way to avoid disaster was for every country to cut its greenhouse-gas emissions. World leaders pledged specific cuts to their emissions for the following 15 years and most countries made an effort to achieve them. Across the world, coal power stations were replaced with natural gas, wind and solar, and electric cars became normal. But after making the easier changes, few countries were prepared to take the harder decisions that would be needed to bring their emissions down to zero. The world's emissions fell sharply, but it wasn't enough to stop the climate becoming much more violent.

When Hurricane Sandy hit New York in 2012, one of the world's mightiest cities was helpless in the face of the onslaught. Confronted with a wall of water, there was no way for the city to avoid the devastation that shut down its transport system, killed 43 people and inflicted at least $19 billion worth of damage.[17] The blacked-out city alternately drowned and burned as the storm triggered fires and left neighborhoods without power for days. New Yorkers were cleaning up for months afterwards.

The good news is that, even with the warming we can expect after this moderate success in cutting emissions, floods like the ones that followed Sandy would still be rare. But they would be much less rare than they are now – and sometimes they would be even worse than Sandy. The 'return period' of a storm is a measure of how long a location should expect to wait before seeing the same kind of event. A storm with a 10-year return period has a 10-per-cent chance of happening in a single year. One study found that the storm surge that Hurricane Sandy caused has a return period of just over 100 years,[18] so at the moment, we might not expect to see another storm like it for the rest of this century. But that is without climate change. New York City's Panel on Climate Change reviewed the evidence about the impacts of global warming on the storms that would hit the city. It found that, in this scenario, floods that currently hit only once a century would strike, on average, every 25 years by the 2080s.[19]

As well as these regular Sandy-level floods, there would occasionally be disasters worse than anything New York has ever seen. The tidal surge that Sandy caused was 4.3 meters above the low-water level; by the 2080s New Yorkers could expect surges another 1.2 meters higher.[20] These floods would only be as common as storms like Sandy were until recently – once in a lifetime events – but if they came, they would make Sandy seem tame. So for New York, even this world of moderate emission cuts is one where Sandy-like floods drown the city around once a generation and where truly catastrophic ones periodically drown the places that manage to escape the more regular floods.

In practice, a city with as much financial value as New York would, when faced with this regular disaster, probably build defenses to reduce the damage. But the entire coastline of the US would be at risk. Some places, like parts of the Gulf Coast, are already familiar with the devastation that storm surges can bring. For them, warming would be likely to herald an even more destructive version of what they are already used to. But in other places, like the northern Pacific coast, rising sea levels would bring flooding so much more frequent than it is now that it would feel like a new threat.[21] In Nova Scotia, southeast Canada, sea-level rise would mean that the kind of flood that is now so extreme it happens only once a century would become a regular event.[22]

While the North American seaboard drowns, Australia would bake. The country is already heating up and, with this level of emissions, it would become ever hotter over the rest of the century. Broadly speaking, the climate would shift around 900 kilometers south, suggesting someone living in Port Augusta, on the south coast, would, by 2090, experience a climate more like the one Alice Springs, in the center, currently has. This means not only higher average temperatures, but many more very hot days. The temperature in Dubbo, a city in New South Wales, currently exceeds 35°C/95°F on about 22 days a year. By 2090, this would treble to 65 days – more than nine weeks each year.[23] At the same time, rainfall patterns are likely to change – in the southwest the amount of rain that falls in winter could shrink by half. Soils would be drier as higher temperatures would mean summer rains would evaporate more quickly. These hotter and drier conditions would be perfect for extreme bushfires.

Not only would these changes directly threaten the health of Australians – with more people at risk from fire and heatstroke – they would devastate farming. As temperatures increased and water became scarcer, it would become ever more difficult to be a farmer in Australia. By the 2070s meat and wool production in some areas could fall by more than 90 per cent – in other words, it could almost cease to exist.[24] Wheat, just ahead of beef as Australia's most valuable farm product, would suffer, particularly as the areas where it is currently most produced are likely to receive less rain at the vital times of year.[25] Some of these problems could be lessened with better irrigation and genetic modification – although both could bring other unanticipated problems – and some areas might become more attractive for farming, but it looks unlikely that this could balance the damage that so much warming would cause. Unless the world sharply cuts its emissions, Australia by the end of the century will be hotter, more ravaged by fires, and will see less rain falling when and where it is needed.

While the world's richest countries struggle against the dramatic effects of climate change, an invisible disaster would unfold in the oceans. Global warming may transform life in the seas – most of the heat trapped by greenhouse gases is absorbed by oceans and, by the end of the century, this scenario would lead to surface waters heating by around two to three times as much as they have over the last 100 years.[26] But warming isn't the only problem for the oceans. Another, less obvious, effect of human emissions may be no less threatening. When carbon dioxide dissolves in water it forms an acid; the seas are now absorbing so much of the gas that they are acidifying at what may be the fastest rate for at least 300 million years.[27]

This double blow may fundamentally change ocean life. One of the first ecosystems to fall would be coral reefs, which would be hit by both the acidification – which prevents them building their structures – and the warming waters, which also weakens them.[28] Even with the warming the world has seen so far, coral reefs seem to be dying. Unless the world limits emissions to the level in this chapter's first scenario, the future for coral reefs is bleak (and even that may not be enough).[29] The loss of reefs would wipe out a vital part of the lifecycle of much aquatic life, but it isn't the only problem. Invertebrate larvae would be particularly sensitive to these changes and many species would struggle to survive.[30] The consequences of collapsing food chains would ripple throughout the oceans. While some species might thrive, the changes would be too fast for most to adapt or evolve and the overall effect would be devastating.

This might seem a side issue. Compared with New York drowning and Australia cooking, why does it matter if the seas get a little warmer and there are fewer fish than there used to be? One reason is practical. Around 10 per cent of the world's population depend on the sea for food and work and would be exposed to these changes.[31] If ocean life is no longer a reliable source of income, the economic collapse, and migrations that could result, may be disastrous. But, beside the practical reasons, a world in which coral reefs – and much else besides – exist only in aquariums and museums would be a sadder and duller place.

The abyss

12 December 2100: The 2015 Paris Agreement had seemed like a turning point in global efforts to avoid dangerous climate change – but it quickly began to unravel. After the US announced its withdrawal, other countries at first insisted that they were still committed to cutting their emissions. Some made a serious effort to expand wind and solar power. But at the same time, many also carried on building more gas power stations and airports, and gave up trying to switch to clean energy. Once it became clear that so many countries were freeloading on the efforts of others, the agreement collapsed entirely and few governments were prepared to try anything that might be expensive or unpopular with the public. It was enough to stop annual emissions growing much further but not enough to reduce them. Emissions in the atmosphere continued to rapidly accumulate and the climate responded.

There are two kinds of changes that would strike the planet if the world did so little to cut emissions. Those living through the transformation would face some changes that are recognizable to the world today, as more extreme versions of those we have already seen in this chapter. But they would also experience some changes that are unlike anything in the world's recent history.

We saw earlier how half-hearted measures to cut emissions would cook Australia; things would be even worse if the world gives up. The city of Darwin, in the north, has always been hot – but its temperature is stable and typically hovers between around 25°C/77°F and 32°C/90°F. That stability saves it from extremes now but it would become a curse if emissions are uncontrolled. By the 2090s, its temperature would exceed 35°C/95°F for nearly three-quarters of the year.[32] Many people would find it unbearable to be outside air-conditioned rooms, and towns and cities like Darwin would start losing their populations as more people moved to escape the heat and fires.

But while the cities of the south, such as Sydney and Melbourne, would still see weather that varied from week to week and wouldn't bake all year long, they would face other threats. One of the most destructive would be cyclones. The evidence on cyclone frequency is mixed and it is possible that the number of cyclones would decrease as the world warms. But it also seems that the cyclones that strike would be stronger and would hit more places – including those further from the tropics – than anything the world now faces.[33] Australia's coastal cities would face being torn apart as they were struck by storms that are unlike anything the region has seen before. Much of the island would become ever more inhospitable. While it is not yet clear how the widening range of storms like these would affect New Zealand/Aotearoa, it appears likely that the country would, overall, also face stronger storms with more extreme winds.[34]

A different kind of threat would hit the UK. Like southern Australia, London and the rest of southern England would face repeated heat waves, every year bringing temperatures that now seem unimaginable. The extreme end of climate models – not likely, but possible – suggests temperatures in London could reach 48°C/118°F with this level of emissions.[35] The intense heat could kill thousands of ill and elderly people. At the same time, crises in other countries would force growing numbers to flee to northern Europe as their lands became uninhabitable. And alongside the dramatic changes, a relentless shift would transform the UK. Even with the warming that would come from these unchecked emissions, the total amount of rain that falls in the country would probably remain unchanged. But the water would arrive in shorter, more intense deluges, meaning more flooding and, perversely, water shortages. For a country renowned for its sogginess, the idea of the UK not having enough water takes some getting used to. Yet in this scenario, the UK – particularly the southeast of England – would face regular shortages.[36] That may not mean the water being cut off from homes, but it would mean rivers and streams across the country running dry and farmers struggling to find enough water for their crops (New Zealand/Aotearoa, whose climate is similar to the UK's, would also face this kind of drying[37]). One day, a child growing up in the UK may wonder why England's land was ever said to be green and pleasant.

Across their vast expanses and climatic zones, Canada and the US would face similar problems to those that will confront people in Australia and the UK. Hurricane Katrina's destruction of New Orleans in 2005 would prove to be an early taste of the ever-more violent storms that would attack the cities of the south and southeast, like Houston, Tampa and Miami. As with Australia, these storms might become less frequent, but those that strike are likely to be more powerful than even the worst to have hit the continent in recorded history.[38]

By the second half of the century, people living in the US's central plains and southwest will almost certainly have faced a drought that lasted 10 years. There is a good chance they will have had to endure one that lasts several decades.[39] This scale of dryness would be at least as bad as the conditions that forced millions to flee when the Great Plains turned into the Dust Bowl in the 1930s, but this time there would be no prospect of the rains returning.[40]

Further north in the US, and in Canada, life would become harder in other ways. As in the earlier scenarios, emissions at this level would bring heat waves that would put the old, young and ill at risk from heart- and breathing-related problems. Earlier, we saw that even the sharpest emissions cuts would mean that Chicago would face heat waves like the one in 1995 every other year by the end of the century. But if emissions rise as much as in this scenario, those heat waves would strike the city three times *each* year. Even in a city that, at the moment, rarely overheats, the summer would bring a constant battle to stay cool.[41] At the same time, the warming would mean increasingly dangerous wildfires, particularly in Canada.[42]

So, across richer countries around the world, this scenario would be tough and for many people it would be devastating. In some places, where the environment now is hard but tolerable, some people would find the changes make life unbearable. Even in the more hospitable, higher-latitude lands, changes would overwhelmingly be for the worse.

But this still doesn't capture the full horror of this scenario. Those changes would be difficult, yes, but many wealthy people in richer countries would find they could adapt and continue a recognizable life, with migration, better water management and air conditioning. To really understand the disastrous consequences of allowing emissions to go unchecked, we need to look not only at these projections – which are fairly well understood and made with a good degree of confidence – but also at some of the impacts that scientists are not yet sure about, but which could be world-changing.

A rising sea level is one of the most widely known and dangerous consequences of climate change, but it is hard to be certain about how much higher the waters will get. This is more to do with the upper limit than the minimum – if emissions don't fall, sea levels would be likely to increase by at least half a meter by the end of this century.[43] Since the sea is slow to absorb energy, that rise would continue for centuries to come. If the world allows emissions to go unchecked, generations would watch the sea rise as a result of decisions made by those living at the moment – by 2300 the seas would be at least three meters higher than they are now,[44] enough to wash away coastal towns, including the homes of over 12 million Americans.[45] Eventually, the seas would reach around nine meters higher than they are now, which is enough to drown the homes of nearly 1 in 10 of the world's current population.

That is the optimistic projection. It gets much worse when we look at the enormous stores of water locked up in ice sheets, whose release would make sea-level rise far greater. Among the most worrying is the West Antarctic Ice Sheet, which appears to be becoming unstable[46] and alone contains enough water to increase sea level by more than three meters.[47] That increase would be in addition to the other expected increases,[48] meaning the oceans could rise higher and faster than the most-used models predict. The collapse of Greenland's ice sheet would be similarly devastating.[49] Researchers don't know how much warming would produce the collapse of each ice sheet and how quickly they could fail. The world might be lucky and find that the ice sheets are relatively insensitive to heat and are slow to disintegrate. But the more emissions humans produce, the greater the chances are that the world won't be lucky and that the sea will climb even further and faster.

Some other consequences of warming may be even more dangerous. As the world heats, there could begin to be vicious cycles that would further increase the rate of climate change. Across the planet, carbon is locked into stores where it is kept safely out of the atmosphere. But that carbon may be released by global warming. The result could be sudden bursts of heat – instead of the slow warming we have seen over the past few decades, the extra greenhouse gases could cause relatively sudden jumps in the planet's temperature.

One store that certainly has enough carbon to trigger disaster is the permafrost in and around the Arctic. These frozen soils contain around 1,700 billion tons of carbon[50] – about 170 times as much as humans emit each year by burning fossil fuels.[51] This permafrost is already melting as the world warms, releasing its carbon in the form of carbon dioxide and methane. That is factored into climate models, but it is possible that the models underestimate the speed at which northern soils may melt and release their carbon. Analysis of the permafrost has found that much of it melted around 400,000 years ago when the Earth was only slightly warmer than it is now – and cooler than the planet would become with this level of emissions[52] (and temperatures appear to be increasing now at a much faster rate than at the end of geologically recent ice ages, giving ecosystems less time to adapt[53]). At the moment, scientists don't know how quickly the permafrost would melt if emissions don't fall. Again, the world could be lucky and find that the soils stay frozen for long enough that they are not an immediate problem. But if it isn't lucky, the carbon could be unlocked at a rate that spins the climate even further beyond what the world can cope with.

Halfway around the planet from the Arctic, another tipping point may arrive. The Amazon rainforest is one of the most important ecosystems on Earth – not only for its diversity of life, but also because of the amount of carbon it removes from the air. Around a quarter of all the carbon dioxide that is absorbed by land-based sources, anywhere in the world, is removed by the Amazon rainforest. What exactly would happen to the rainforest if emissions don't fall is still unknown. The Amazon is extraordinarily complex and the plants in it would probably respond in different ways to changes in temperature and rainfall. It is possible that much of it may be able to survive even fairly extreme warming, at least for a few decades.[54] Or, global warming might mean that the climate is no longer suitable for a tropical rainforest, and the ecosystem could collapse.[55] Whether the forest went with a conflagration that burnt millions of hectares, or slowly died, the result would be the same. The planet would have lost a section of its lungs and humanity would have taken another step into the abyss.

The choice

We have three visions of the world as it will be during the lifetime of children alive today. One, where the world gets serious about cutting emissions and limits warming to not much more than the amount there has already been. Life would be harder for many people, particularly those in poorer countries where droughts and changes in food prices can spell disaster, but, by and large, this is a world that is not so different from the one we know. This is the scenario that is widely described as 'safe'.

The second, messier scenario, is where most countries make a moderate effort to cut their emissions, but warming still goes beyond the limit that countries agreed to aim for at the Paris climate conference. Storms would ravage and flood coastal areas, heat waves would scorch places that were once bearable, oceans' food chains would collapse and hundreds of millions would find they have no choice but to flee their homes. This book refers to this as 'dangerous warming'.

The third scenario – where the world's attempts to cut emissions are no more than half-hearted – takes us to a place of nightmares. If the world is lucky, it would have to live with nothing worse than droughts and floods, heat waves and storms of astonishing ferocity, and mass migrations triggered by the collapse of food production. If it is not lucky, humans would unwittingly unleash a store of buried carbon, tipping the world beyond the worst of any of these projections. There would be no going back to the world we know – a scenario that can reasonably be described as 'catastrophic' or 'disastrous'.

Part 2

The causes of apathy

5
Sight and mind

Climate change is rarely mentioned in everyday life. Even when it is talked about, apathy is usually not challenged, as most people are inclined to dismiss threats that appear complex, distant and slow-moving. But this link between psychology and climate apathy is not inevitable.

'Never put off till tomorrow what you can do the day after tomorrow.'
– Mark Twain

In Part 1, we saw that climate apathy is widespread in many of the countries that are most responsible for global warming. This apathy is less visible than climate denial, yet it presents a fundamental threat to efforts to limit climate change. Over the coming decades, many people would have to change aspects of their lives if the world is to limit emissions enough to avoid dangerous warming, but unless climate apathy is addressed, there won't be much support for such measures. Governments and businesses will be tempted to take only the easy decisions, endlessly putting off the hard ones. If that doesn't change, the warming could be devastating.

The rest of the book turns to the question of how apathy can be beaten. Part 2 investigates apathy's causes – it looks at how climate change is talked about and why it isn't, and how it is often described in ways that make it seem unimportant. Part 3 explores what can be done to influence how people who are apathetic about climate change think, talk and act about the problem.

To begin, this chapter looks at how far climate change features in most people's day-to-day lives – in the news, TV and films, and in everyday conversations. As we will see, it is largely absent from daily life, and this is one factor behind apathy. But that isn't the only factor.

The chapter also examines how the nature of climate change interacts with human psychology to make the problem feel unimportant.

Often forgotten

Like most people who are worried about climate change, I'm often frustrated by how little attention the media pay to it. Climate change seems only rarely to attract headlines, while news on other topics that are closely related to it – like airport expansion or shale-gas drilling – usually fails to mention it. Even coverage of unusual weather, such as unseasonal heat waves, doesn't often make a link. For an issue that is one of the most important facing the world and will only be addressed with changes to most people's lives, it doesn't seem to attract nearly as much coverage as it might be expected to.

But that is just my feeling. Perhaps those of us worried about global warming are being too demanding in our expectation that it should get more coverage. Can we trust our feeling that climate change is under-reported?

The extent of media coverage of climate change has varied in recent years. Research by the University of Colorado, which monitors international newspaper coverage of climate change, shows it was mentioned increasingly often until late 2009. But since then coverage has fallen. As we saw in Chapter 2, concern about the problem dipped in some places after 2009, following the Copenhagen climate conference. Media coverage did the same, although since 2015 coverage has generally begun to increase again.[1] As you would expect, the peaks in coverage coincide with events that prompt journalists to write about it. My research of coverage of different kinds of these 'triggering events' found that extreme weather, like flooding, sometimes leads to an increase in media mentions of climate change, but international climate conferences and major reports from the Intergovernmental Panel on Climate Change (IPCC) are more consistent in generating coverage.[2]

But even when climate change coverage increases, it still seems to be dwarfed by the attention given to other topics. A UK study found that there was about the same coverage of health news in just one month of 2006 as there was of climate change in the whole of the previous year.[3] Of course, health covers a range of sub-topics, in the way climate change might itself be considered a sub-topic of the environment, so perhaps this isn't a fair comparison. But the research also found that stories about the climate were outweighed by stories about crime and the economy, and that coverage of possible ways of addressing global warming, such as emissions trading, tended to be relegated to the business pages. The study showed that climate coverage was mostly confined to the left-leaning 'quality' newspapers, and was rarely mentioned in the newspapers that, at the time, had higher circulations. More recent research has found the same.[4] The growth of newspaper websites has transformed these differences in circulation, so receiving coverage only in left-leaning quality newspapers is no longer a route to obscurity. Nevertheless, if we believe that climate change is an issue of comparable significance to crime, health or even the economy, we have to conclude that coverage of it is sparse.

Other studies have looked at not just how much the media talk about climate change, but at what they say about it. A 2007 study found few articles in quality newspapers in the UK and US that disputed mainstream climate science,[5] but a follow-up study found that tabloid newspapers were significantly more likely to challenge the evidence.[6] Analysis of coverage of the IPCC's reports similarly found that most of the UK and US stories focused on ideological disputes and debates about whether climate change was real. The coverage gave less attention to the likely consequences of global warming or the opportunities and benefits of tackling it.[7]

Research on climate change in film and TV has similarly concluded that the issue is largely ignored. Analysis of a database of 87,000 TV and film scripts found a roughly five-fold increase in mentions of climate change between 2004 and 2009 (notice how that same year keeps coming up). But after that, the references dropped sharply, by 2014 falling back to near 2004 levels.[8] Even during those years when climate change appeared more often in films and TV, it never became a part of the furniture. The 2004 disaster film, *The Day After Tomorrow*, Al Gore's 2006 film and book, *An Inconvenient Truth*, and Leonardo DiCaprio's film of the following year, *The 11th Hour*, were all widely seen and talked about. Yet they were all explicitly *about* climate change – none sought to tell other stories with climate change providing a background. The same is true of DiCaprio's 2016 film, *Before the Flood*. Literature has made more efforts to use climate change as context for other stories. But, while some well-known writers, such as Margaret Atwood, Barbara Kingsolver, Ian McEwan and Kim Stanley Robinson, have incorporated climate change into their novels, and there is now a genre of climate-change stories, cli-fi, the subject still appears in only a vanishingly small minority of books, as it seems to throughout popular culture.

Many people hear about climate change in school; it is part of the school curriculum in many countries. But most people aren't children, so even widespread climate-change lessons would still only reach a small part of the population. There are also regular arguments about the accuracy of explanations of global warming in textbooks, which often seem to exaggerate scientific doubts. Analysis of Californian sixth-grade science textbooks in 2015, for example, found that climate scientists were wrongly presented as being split on the basic causes of climate change.[9] Similar concerns have been raised about textbooks in other US states and countries.

Why does it matter that most people don't hear much about climate change? We shouldn't take it for granted that public opinion is determined by how an issue is covered in the news, how it appears in films, books and TV shows, or what children are taught at school. Yet social scientists generally do believe that media coverage has at least some effect on public opinion – although the nature of the relationship is disputed. While some suggest that media coverage directly influences public opinion by making people more interested in a topic,[10] few claim that the relationship between news coverage and public opinion is predictable or consistent. Many researchers say that responses vary across different kinds of media and groups within the public.[11]

There is a model of how people form opinions that can be helpful for understanding why it matters what people hear about an issue like climate change. John Zaller, a US political scientist, suggested that, when an individual hears an argument about any subject, they might accept the argument or they might not.[12] If they do accept the argument, they store it in their minds along with other information on the subject, ready to be selected from when they are prompted to consider what they think. More engaged individuals are likely to hear more arguments but tend to be more discerning about which they accept, rejecting arguments they recognize as being in conflict with their existing values. This means that people who are less politically aware – but still engaged enough to hear some arguments – are more likely to hold conflicting opinions. An effect of this is that many people don't have a consistent or fixed opinion on a subject – what they say about it depends on when they are asked and how the question is phrased.

In Zaller's model, arguments derive from elites and are channelled through the media. The rise of social media and more diverse sources of news and opinion may mean a small class of elites have less influence over public opinion than they used to, but the theory still seems reasonable. An individual receives information, which they may or may not accept, depending on their other beliefs, and then they sample from the relevant arguments in their mind when prompted to do so.[13] The model, which Zaller summarized as Receive-Accept-Sample, seems to fit the evidence that public opinion about climate change often appears contradictory. It suggests that media coverage can influence public opinion, but that such influence isn't inevitable or straightforward. Later chapters, which explore how we can influence opinion about the climate, will draw on this approach.

But the relationship between the media and public opinion clearly isn't one-way. Newspapers need readers; TV shows and films need viewers. Even if they can shape public opinion, they struggle if they cover subjects that few people are interested in. This is reflected by the fact that, when there is more coverage of climate change, the public are more worried about it.[14] That might indicate that media coverage increases public interest, but it seems at least as plausible that the media cover stories that their audiences are interested in. Public opinion and news coverage are locked together – public interest can encourage more coverage of a topic, which can drive further public interest.

As well as being mostly absent from the media and popular culture, climate change isn't a regular topic of conversation for most people. This may be because most people aren't often prompted to think about it. Not only does it not feature much in news, TV, film and books, but most people rarely experience climate change in day-to-day life in obvious ways. Perhaps as a result, nearly 60 per cent of Americans say they never, or only rarely, talk about climate change with their family and friends.[15] If we exclude the base segment (the *Alarmed*), who don't need much more persuasion to worry about climate change, that rises to 70 per cent who rarely, or never, talk about it. This is sometimes labelled the climate silence.[16]

As a result, climate change is often left out of conversations where it should be central. This is illustrated in a long-running debate about whether a new runway should be built at a London airport and, if so, which airport should be expanded. The main argument used in favor of expansion is the apparent economic benefit, while the chief argument against it is the disruption expansion would cause to local communities. The climate is rarely mentioned. In fact, those who favor expansion have managed a remarkable linguistic trick to define the 'environment' to mean anything other than the climate. One of the airports hoping to be the site of a new runway claims to offer an environmentally friendly option because the increased noise would affect fewer people than its competitors and the airport would supply noise insulation for affected residents.[17] Even though building a new runway would make it much more difficult for the UK to hit its legal emissions targets,[18] the debate about expansion has largely ignored emissions. Since climate change is such a small part of everyday life for most people, decisions like this, which will influence whether the world can avoid dangerous warming, are taken with almost no discussion of the climate.

Is it news?

A different looming global catastrophe was, not so long ago, a favorite topic of the media. As 1 January 2000 approached, increasingly alarmed media coverage warned about a coming meltdown. It wasn't the icecaps that were predicted to collapse, but global computer systems. The Millennium Bug would, apparently, result in computers being unable to cope with the year changing from 1999 to 2000, and was said by tech experts to put at risk every system that used dates, from banks to air-traffic controls. During the course of the year leading up to the fateful date, UK newspapers ran on average over five articles a day mentioning the Bug – more than twice the number that mentioned climate change.[19] Given the attention, it is not surprising that more than half of the public thought the Millennium Bug posed a serious threat to services.[20]

The world didn't end on the stroke of midnight. There were some problems with computer systems, but even a catalogue of the global errors caused by the Bug lists problems no more disastrous than incorrect medical information being sent to pregnant women – an awful situation for a few people but nothing close to the predictions of global doom.[21] There is still disagreement about whether the problem was always vastly exaggerated or if it would have been worse if so much work hadn't been put into resolving the problem. But it is clear that, in comparison with the dangers that uncontrolled emissions would cause, the Bug was never going to be more than a blip. And yet before it fizzled out, it was a favorite of the media and became so widely known about that the US government was worried that the public would start stockpiling food.[22] Given the media interest in this other threatened catastrophe, why is there so little media coverage of climate change?

In many ways, climate change makes for bad news. Not bad in the sense of 'uh-oh' (although that is certainly true) but bad in the sense of 'doesn't interest people'. There are a few reasons for this. One is that global warming doesn't easily offer the classic ingredients of news stories. It is not that it is missing absolutely everything journalists look for. A common feature of a news story is a threat, such as a previously unknown danger, a deteriorating situation, or a growing menace. Climate change clearly offers this but it often struggles to provide another vital ingredient. Most news stories contain, as well as a threat, an element of conflict. This can come in the form of a physical confrontation; sometimes it is in disagreements about an issue, for example when a story is about government policy. This demand for conflict in news stories shapes how climate change is covered.

Sometimes climate change can easily provide conflict. That is particularly true for protests against plans to extract and use more fossil fuels, such as attempts to drill in the Arctic or the Great Australian Bight, to build a pipeline or to frack for shale gas. There is no difficulty in finding conflict in these stories, where the news can focus on the people opposing the operation. But even in these stories there can be a problem with the way the threat is described. Coverage tends to focus on the visible and tangible dangers of local pollution from, for example, an oil spill or drinking-water contamination, rather than the longer-term and more complicated threat of climate change from burning fossil fuels. This lowers the bar for those who want to build the infrastructure. To win the argument, they only have to be seen to have addressed concerns about local safety or sufficiently compensate anyone who might be affected. They can even argue that the national interest justifies the suffering of a few local people. These might not be easy arguments to win, but they are not as hard as showing that extracting more fossil fuels is compatible with avoiding dangerous global warming. This is a reason supporters of airport expansion prefer to ignore climate change and instead focus on local environmental issues, which they may be able to do something about.

The media can also create an impression of conflict by focusing on disputes about climate science. Such is the appeal of this kind of story that journalists sometimes selectively quote from new academic research to give the impression that it indicates that climate change is less threatening (or, sometimes, more threatening) than the scientific consensus suggests, even when the authors insist it does no such thing.[23] Despite clarifications from the researchers whose work is misrepresented, some journalists continue to find this an easy way to cover climate change. And even when journalists try to avoid misrepresenting scientific findings like this, they often instead find conflict by focusing on apparent public doubts about the reality of global warming.

A second problem with climate change as a news story is that, from an outsider's perspective, it doesn't seem to develop much. News stories are, as the name implies, about things that are new: events, dramatic changes, or discoveries of something that had previously been secret. The media generally don't deal well with slow, long-term changes, or with gradual improvements in scientific understanding of the world, however important the stakes.

This spells trouble for coverage of climate change. Understanding of the issue is sufficiently developed that there are now few dramatic breakthroughs that transform knowledge on major aspects of it. When advances are made in understanding the causes and consequences of climate change, many scientists are reluctant to describe the research in the kind of unequivocal terms that make for powerful news stories. They, understandably, tend to emphasize uncertainty in the findings. Inevitably, the media are drawn to researchers and campaigners who are prepared to emphasize outlying projections and downplay the related uncertainties, including projections that overstate the risk of disaster as well as ones that understate it.[24]

So, journalists are most interested in stories that show threat and conflict and where there is dramatic change. This is a problem for coverage of global warming, where the most easily available conflict is in supposed disputes among scientists, and where it is relatively rare for there to be scientific breakthroughs that seem important to outsiders. Climate change offers journalists a story with a compelling threat if they focus on its consequences, but it can still be hard to show dramatic developments in understanding of its likely impact.

One other factor that might reduce coverage of global warming is the influence of advertising and ownership of the media. It is possible that businesses with a strong interest in downplaying the threat of climate change – particularly those with stakes in high-emitting activities, like fossil-fuel companies and airlines – might threaten to withdraw advertising from media organizations that focus on climate change.

There is some evidence that news organizations do at times avoid covering particular stories that could damage their advertisers. In 2015, a leading journalist resigned from the UK's *Daily Telegraph*, accusing the paper of holding back coverage of a banking scandal to protect HSBC, a major advertiser (an accusation the paper denies). It is hard to come by evidence of how widespread this is, but a study in 2000 of US journalists reinforces the suspicion that it does happen. According to the research, only 27 per cent of journalists say they never avoid stories that could damage advertisers; 29 per cent say it happens sometimes or is commonplace.[25] It also found that 77 per cent of journalists at least sometimes avoid stories that audiences might find 'important but dull'.

It's not you, it's us

So far we have seen part of the explanation for why there is so little discussion of climate change: it doesn't offer the classic ingredients journalists look for in news stories and it sometimes conflicts with media organizations' commercial interests. This lack of coverage keeps it out of the public eye. But that doesn't seem like the full answer. Newspapers, commercial websites and TV channels only survive when they have people paying for them or providing an audience for their advertisers. If they don't cover climate change, perhaps that is because they think their audience isn't interested.

If so, maybe they have a point. In recent years, a number of studies have explored how the human mind tends to stop people worrying about climate change, and have found that there are several aspects of climate change that incline most people to avoid thinking about it.

One of the greatest difficulties the mind has with climate change is its complexity. This is more about the consequences of global warming than its causes. The latter aren't hard to understand – particular gases, which humans produce when burning things like coal, oil and gas, insulate the planet like a blanket and so heat it up. But the consequences of this are much more complex. Subject to different levels of warming and different timeframes, some places will get drier and others will get wetter; some will experience floods and others droughts (often the same place will face both); disruptions to weather patterns might mean that some places will face colder winters, despite the planet warming up overall.

This complexity about the consequences of climate change makes it harder for people quickly to see why it matters. Unlike with the ozone hole, which has a straightforward impact – skin cancer – it is relatively hard for people with only a passing interest in global warming to grasp its likely consequences. This is exacerbated by the language many people worried about the problem often use to describe it, as we will see in the next chapter.

A further difficulty for human psychology is climate change's 'futureness'. In general, the human mind avoids devoting effort to things that will happen in some far-off year, however scary they might be. Through evolution, humans have come to be excellent at dealing with threats that are direct and visible. As crises like the Second World War demonstrated, humans can make radical changes to their lifestyles for a perceived common benefit and, in such extreme circumstances, many people are prepared to sacrifice their own lives to combat society-wide threats. But this isn't a clear parallel. The Second World War involved an obvious and dramatic threat – which was existential for some countries – that clearly demanded an extreme response. When threats aren't clear or immediate, they don't capture attention well. Daniel Gilbert, a Harvard University professor of psychology, has described the brain as a 'get-out-of-the-way machine'[26] – good at avoiding things that are visible, but not so adept at dealing with distant threats. Climate change is one such complicated and (mostly) distant danger, but not the only one. From health hazards such as antibiotic resistance to environmental problems like the build-up of plastic in the oceans, many people have a sense of global threats that may someday trigger enormous problems, but most still pay them little attention.

Attention to climate change is also hindered by the way it advances quite slowly. Describing global warming as slow-moving might seem strange when the planet is beginning to heat at a rate that appears unprecedented – the current release of carbon may be the fastest since at least the time of the dinosaurs[27] – and the warming is already leading to dangerous weather extremes. And yet the timescales might still be too slow to spur many people to act. Over just a few years, the mind starts considering new weather patterns to be normal, forgetting how different they are from what used to be typical. In the context of biodiversity loss, the marine biologist Daniel Pauly labelled this 'shifting baseline syndrome'.[28]

But even more than for any of those other threats, many people in richer countries are tempted to put climate change out of their minds because it seems like a problem that will not affect them directly. When, in many of the richer countries with high emissions, people worried about climate change describe the problems that dangerous warming would cause, the focus tends to be on two sets of other people: those living in poorer countries, and those who will be alive in the future. Since human psychology attunes people to threats that are immediate and directly affect themselves and the people closest to them, this representation of climate change as a distant threat offers another excuse to put off paying attention to it.

A further reason many people don't often think about climate change may be that it doesn't easily fit into the established storylines the human mind expects the world to follow. Just as journalists look for stories where they can show threats, change and conflict, the mind tends to interpret what it hears about the world through well-established storylines. The tales we read, hear or watch are, arguably, all variations on one of a small number of familiar stories, such as the underdog triumphing against the odds in a battle of good versus evil (as in *Star Wars*, *Terminator* or *The Hunger Games*).

The trouble with climate change, as George Marshall pointed out in his 2014 book, *Don't Even Think About It*, is that it is hard to fit it into any of these storylines. Some people attempt to present global warming as a struggle of good against evil, with scientists and activists battling fossil-fuel companies and others who deny its reality. But this doesn't hold up to much scrutiny. While energy companies extract and transport oil, gas and coal, it is ordinary people who use the fuel and electricity that they provide. Embracing a good-versus-evil storyline would require most people to think of themselves not just as good people doing bad things, but as the baddies at the heart of the story. A more palatable alternative may be for global warming to instead be presented as a story of rebirth, as in Dickens' *A Christmas Carol*. But even so, people would first have to be persuaded of the error of their ways before they accept they should change. The difficulty in applying any of these storylines is that, to most people, climate change isn't self-evidently the consequence of an immoral act. Unlike some cartoon villain, the people who cause climate change – nearly everyone – aren't obviously evil. Thinking of ourselves as the ones at fault is still an enormous leap.

Another reason human psychology may incline most people to ignore global warming relates to the kind of losses associated with it. Studies have repeatedly found that most people are determined to avoid losses, even if that means failing to secure gains. For example, research in 1984 by Daniel Kahneman and Amos Tversky showed how focused most people are on avoiding losses.[29] In one study, they tested participants' responses to two notional vaccine programs, designed to address an imagined disease outbreak that would, unless stopped, kill 600 people. Participants were presented with two alternatives:

> If Program A is adopted, 400 people will die.
>
> If Program B is adopted, there is a one-third probability that nobody will die and a two-thirds probability that 600 people will die.

Even though the average outcome of both programs is the same – that 400 people will die – 78 per cent chose Program B, gambling that they might reduce the deaths to zero. But perhaps that wasn't loss aversion – maybe participants were mostly attracted by the possible positive outcome of zero deaths, rather than repelled by the guaranteed loss, in Program A, of 400 lives.

In a neat twist, which suggested that it was the desire to avoid a guaranteed loss that drove the result, the researchers also tested a differently worded description of the vaccine programs, this time phrased in terms of lives saved rather than lives lost:

> If Program A is adopted, 200 people will be saved.
>
> If Program B is adopted, there is a one-third probability that 600 people will be saved and a two-thirds probability that no people will be saved.

The options are mathematically identical to those in the first pair – if participants were behaving rationally they should have picked the same option. But this time the responses were reversed with 72 per cent choosing Program A (I feel the same: despite knowing there is no difference between the descriptions, my mind is screaming at me to choose Program A this time). The driving force seems to be loss aversion. With the first pair, participants don't want to condemn 400 people to die, and in the second they don't want the risk of losing the 200 lives they could definitely save. In this experiment, the options all had the same average outcomes so loss aversion didn't lead participants to make decisions that brought them costs, but other studies have shown that people are often unwilling to accept a net benefit if it would mean sacrificing what they already have.[30]

You might think loss aversion would make most people determined to avoid the losses that extreme global warming would bring, but it depends on what is seen as the losses at stake – what is the sacrifice and what is the potential benefit? A good example is in the possible trade-off between limiting climate change and promoting economic growth. If most people consider a stable climate to be something they already have and future economic growth to be something they might gain, they might be expected to prioritize protecting the climate. But few people consciously think of the stable climate as an asset that they could lose. It seems easier for most people to imagine risks to an economy that is expected to grow and deliver prosperity.[31] The powerful mechanism of loss aversion swings into action against anything that seems to jeopardize the economy, even if that unwittingly means allowing losses to assets that are less obvious, like the stability of the climate. It is hard to imagine that, over the next few years, most people will come to see the economy as a lower priority than a stable climate, which is one reason why it will be difficult to win widespread support for measures to limit climate change if it is widely believed that such measures could only come at significant cost to the economy.

There is one final factor in the relationship between psychology and climate change, which not only discourages many people from paying attention to the problem, but also makes it more difficult to reach consensus on how to deal with it. Unpleasant though it may be for those with a liberal mindset, most people split the world up into what social psychologists call 'ingroups' and 'outgroups' – people they identify with and those they don't. The groups are divided on the basis of characteristics such as ethnicity, religion, social class and political beliefs. People tend to favor others in their ingroup in all kinds of ways, including being more sympathetic to their opinions, while tending to be more critical of the views of people in outgroups. Attachment to the ingroup is so strong that, when people hear evidence on a topic that relates to their ingroup's identity, they typically respond by strengthening their existing views, even if the evidence in fact contradicts what they believe.[32]

Climate change is a victim of this split. As George Marshall shows, disputing climate science is central to some groups' identity, notably among some US conservatives.[33] He suggests this is, at least in part, a product of the way climate change is presented by opponents of government intervention as an excuse by the political Left to justify higher taxes and more regulation. The problem with climate change being associated with any group's identity is that it makes it nearly impossible for an outsider to influence the opinion of members of the group. Scientific evidence is a feeble weapon against identity.

Group identity is one of several psychological factors behind climate apathy that might feel beyond our control. The consequences of climate change are complex and slow moving and the most extreme impacts will be felt only in the future, so many people are tempted to put it out of their minds. It doesn't easily fit into the story templates that the human mind uses to understand the world, and the losses associated with it may not be the ones that most people are conscious of and seek to avoid.

But, though we can't change human nature, we can change how global warming is talked about and understood. Some of the ways in which apathy is worsened by human psychology are the avoidable result of how climate change, and the solutions to it, are described. There is nothing inevitable about the examples that come to mind when most people think about the consequences of climate change, the stories the media tell about it or the way it fits into existing divisions between ingroups and outgroups. These are the consequence of choices made by people who talk about global warming. The next two chapters explore how some of the ways in which climate change is described, often designed to increase attention to it, and concern about it, have in fact fostered apathy.

6

Nothing to worry about

Apathy is worsened by the way climate change is often discussed by people worried about the problem. Descriptions of climate change often focus on distant places and animals and use apparently unthreatening small numbers to describe average global-temperature change and annual sea-level rise.

'I am not a climate-change denier... I accept that the globe is warming mildly, 0.8 of 1 degree in 131 years.'
– Lord Donoughue, Member of the UK House of Lords[1]

We have seen that climate change isn't big news. Despite the threat it poses, it doesn't feature much in films, TV or books, nor is it a popular conversation topic. While some people are enthusiastic about tackling it and a few are enthusiastic about arguing with those people, most are apathetic. This apathy is partly caused by the complexity, distance and slowness of climate change, which makes it badly suited to attracting most people's attention. But the nature of climate change is only part of the problem – at least as important is how it is described.

This chapter begins to examine how this causes climate apathy. It shows how the language and arguments that are often used to warn about global warming put many people off paying attention. Climate change is just one issue of many competing for swings' notice. Yet, instead of emphasizing the arguments that could interest them, discussions of global warming often focus on aspects that appeal mostly to people who are already worried about it. At the same time, the terms used to explain the threat it poses are often misleadingly reassuring to anyone who isn't already an expert. The result is that climate change often sounds, to many of the people whom we need to win over, like it is nothing to worry about.

In the frame

Three advertisements, from different campaigning groups, reflect how climate change is often presented.

The first video, for a campaign against flying, is set among skyscrapers, with the viewer looking up from ground level. We see objects falling from the sky and after a few seconds realize they are polar bears. As they fall, they crash horribly into the buildings and road with an awful thud, crushing a car and spraying blood. The closing text reads: 'An average European flight produces over 400 kg of greenhouse gases for every passenger... that's the weight of an adult polar bear.'[2]

The second video also focuses on the damage climate change would cause to nature. It starts with a chimpanzee in the burnt remains of a forest, then switches to a polar bear staring at an iceless sea, and finally a kangaroo in a desert, beside a railway track and an industrial plant. We return to each in turn, seeing them kill themselves – the chimpanzee by hanging; the bear by flopping into the sea; and the kangaroo, gruesomely, by jumping in front of a train. The closing text tells us: 'If you give up, they give up... stop global warming.'[3]

An online advertisement, published at Christmastime, shows Santa Claus' hat floating in the seas of a melting Arctic. The text says simply: 'Save Santa's home.'[4]

These are not the only ways campaign groups talk about climate change, but they are not unusual. When people worried about climate change talk about the threat that it poses, their focus is often on the damage it would cause to distant places and animals. Yet such warnings are unlikely to persuade the swings that they should care about climate change.

To understand why this is, and what could be more effective instead, we need to consider how the brain processes information about complex issues. George Lakoff, a professor of cognitive science linguistics at Berkeley, has probably done more than anyone in recent years to improve understanding of how people think about climate change. The central lesson from his work is that facts alone don't win arguments. What matters most is how information triggers 'frames' – established pathways in the brain that are associated with certain meanings and emotional reactions. Frames are built up over long periods through frequent repetition. Once an issue is associated with a particular frame, it can't be quickly adjusted to trigger different concepts. As Lakoff points out with reference to many examples – such as the framing of tax as something that one can have 'relief' from, or the repeated description of Trump's 2016 opponent as 'Crooked Hillary' – conservatives have generally been better than liberals at establishing and reinforcing frames that encourage people to see the world as they do.[5]

Advertisements like the three described above frame climate change as a problem for animals and distant places. Since frames can only be established with considerable effort and cannot easily be changed, it is important to consider how framing climate change in this way is likely to influence how the swings perceive the threat.

Everyone with compassion cares about disasters that cause suffering and loss of life, wherever in the world they happen. This is obvious from the way such disasters can mobilize vast numbers of people around the world to try to help. Crises like the 1984 Ethiopian famine, the 2004 Indian Ocean tsunami and the 2015 Nepal earthquake prompted many to donate.

But scratch the surface and these responses to foreign disasters seem discouraging as models for action on climate change. Public engagement with the crisis is usually brief. The response of most people who take action is limited to giving money – an important gesture, but one that takes little time or effort. Far fewer people make deeper sacrifices. What's more, after the immediate response and coverage, the disaster is forgotten. Ask yourself: how often in the past year have you thought about the 2010 earthquake that killed around 220,000 people in Haiti, or the floods of the same year that left about a fifth of Pakistan under water? If you are like me, the embarrassed answer is probably somewhere between 'not often' and 'not at all'. I think of myself as an internationalist, yet I can more easily bring to mind disasters that killed fewer than 100 people in my own country than I can recall far larger disasters in other countries. Disasters happening in far-off places may attract compassion in the moment they happen, but they rarely move most people to do more than give money and they are quickly forgotten by most who aren't affected.

The same applies to disasters that happen to distant places where very few people live, such as the poles. There are good reasons for people worried about climate change to pay attention to the Arctic and Antarctic. Melting ice is bad news for many reasons. It is an early warning of changes to the global climate, an accelerator of further warming (the darker land and ocean revealed by melted ice absorb more heat) and a direct threat in itself as increased water means not only higher sea levels[6] but, partly because the melted ice is not salty, also likely changes to ocean circulation and therefore to weather patterns. Yet, in themselves, the poles have little emotional power for most people who are apathetic about climate change. Few people have been there, still fewer have lived there and they rarely feature in popular culture. What is more, you have to be knowledgeable about climate change to recognize why we should be particularly worried about the melting of the poles. If you don't already know about the effects of ice melt on sea levels, ocean circulation and accelerated warming, it is unlikely you would be especially alarmed by a message about Santa's home being threatened. It seems unlikely that warnings about the Arctic melting would have a great deal of impact on swings like the *Cautious*, most of whom say they are not very well informed about climate change.[7]

This is also true of the impact of climate change on animals. Many, perhaps most, people describe themselves as animal lovers, but this love doesn't seem unconditional. Every so often an investigation breaks a scandal about the treatment of animals raised for food. Most people know that much of the meat eaten in richer countries comes from animals that lived and died in terrible conditions. Even countries with relatively tough animal-welfare standards still often put animals through brutal ordeals.[8] Yet while some people are moved by this knowledge to buy meat that claims to be ethically produced – or to cut down or give up meat altogether – most continue eating meat exactly as before. There isn't much reliable evidence on the numbers who don't eat meat, but polls in some richer countries suggest that only around two to eight per cent are vegetarian, with little sign of this increasing.[9] It seems that rates of meat consumption are linked with economic trends and generally increase during periods of growth.[10] For all the shocking evidence of how animals are treated for the plate, which conflicts with the values many claim to hold, most people don't seem to be changing their eating habits.

Some people are indeed motivated by the plight of animals – but these are most likely to be people who are already worried about climate change. For example, 87 per cent of Americans in the base *Alarmed* segment said they consider themselves to be environmentalists.[11] In Australia, that segment is easily the most likely to say they have a connection to nature.[12] But the US study also found that few of the swings closely follow environmental news – only 39 per cent of the first swing segment, the *Concerned*, and 14 per cent of the second swing group, the *Cautious*, say they do.[13] The equivalent segments in Australia are much less likely to say they are connected to nature.[14]

Advertisements like the ones described above appear to be targeted at the base segment. There is nothing inherently wrong with this – all campaigns need to make sure their strongest supporters keep donating and volunteering, and organizations whose existence depends on their membership have to keep reflecting their members' interest. But if everyone worried about climate change talks only to those who already agree with them about it, there won't be enough support for the measures that will be essential for avoiding disaster. Even if every member of the first swing group, the *Concerned*, could be won over and added to everyone in the base segment – already a tall order – we would still have only around half of the population, which would not be enough for some of the changes that are likely to be needed. Broader support will also depend on winning over at least the *Cautious*. To many people like these, the description of climate change as a threat that only, or even just mostly, harms distant places and animals might not only fail to challenge climate apathy – it might worsen it.

If we are to persuade swings to think differently about climate change, we have to be aware of what is going wrong with how the issue is currently framed. If the term 'climate change' activates a frame that does nothing to persuade most swings that it matters to them, their apathy will remain unshaken. While some people are motivated by threats to distant places and animals, most are not. The framing of climate change as a problem for polar bears and melting icebergs certainly motivates some people to take it seriously. This may have mobilized the activists who first drew attention to the problem and are still essential for pressuring the worst polluters. But the framing is ineffective with the people whose support will be increasingly important.

Fortunately, there is some evidence of what can work instead. Several studies have looked at how personal experience of extreme weather influences views of climate change. One set of UK studies has consistently found that people who have been the victims of floods linked with unusual weather are more likely to be worried about climate change and to consider it a serious threat to themselves and their family.[15] Research in the US also found that talking about how climate change may affect the health of Americans, such as with poor air quality and increased risk of disease, generated positive responses. While this effect was strongest in the segments that are already most worried about climate change, the health angle was also received positively by those in the swing groups.[16] Other studies have not always found the same – for example, Australians affected by natural disasters don't seem to be more worried about climate change, although those who describe their experience as the consequence of climate change are indeed more worried about it[17] – but the overall picture seems to be that focusing on the local consequences of climate change has a good chance of influencing those who are currently apathetic.

This is reflected in responses to a few disasters that happened in far-off places, but which didn't follow the pattern of the ones described earlier. After the Ebola virus was identified in Guinea in March 2014, the crisis attracted an impressive level of media and public interest in countries that were far from the outbreak. In the second half of 2014, when coverage of the virus peaked, the story attracted more Google searches in the US than the high-profile protests in Ferguson, Missouri.[18] Over the first year of the crisis, the virus killed 10,460 people.[19] That's usually not enough deaths of people in poor countries to attract such public interest in rich ones – malaria killed 429,000 people in 2015[20] and more than 5,000 died in violence in the Central African Republic in the first nine months of 2014,[21] yet neither came close to being a news story on the same scale. So what made Ebola such a global story?

The explanation is simple: Ebola had the potential to affect people in rich countries who are not threatened by distant violence or tropical diseases like malaria. When disasters strike people in one country, but have the potential to directly affect those who live far away, people in those other places are much more likely to pay attention. The recent electoral success of nationalism has rested on appeals to self-interest. While many people resist such appeals – and limiting climate change depends on challenging isolationism – it is clear that the local effects of global threats have the most emotional impact. Many people may see this as a bleak statement about human nature, but it is one we can't ignore when we are thinking about climate apathy. While people in poor countries are likely to suffer the most from climate change, we can't assume that worry about these victims will be enough to motivate a majority of people in rich countries to support action.

Given this, it is clear that there is a particular problem with the world's most high-profile effort to show why climate change matters. Every few years, the UN's climate science body, the IPCC, summarizes current knowledge about global warming into several lengthy reports. The publication of these reports is always a major event, but the way they are presented consistently misses the chance to frame climate change as an issue that matters to the swings.

For news about climate change, the reports receive an unusual level of coverage – for example, the September 2013 report prompted more mentions of the issue in UK newspapers than any other event for nearly two years[22] – yet they had little direct impact on public opinion. As we saw in Chapter 5, most of the coverage of the 2013 report was focused not on the likely consequences of global warming, but on arguments about whether the science was credible. After it was published, the proportion in the UK who said they considered the environment one of the top issues facing the country barely moved, increasing from four per cent to five per cent.[23] In Australia, the report did coincide with an increase in the numbers identifying climate change as a key issue, from four to nine per cent,[24] though this may have been to do with the bushfires that were burning at the time. Both compared favorably with the US, though, where the proportion picking the environment as a top issue barely registered, at one to two per cent.[25]

Behind this lack of public response to the IPCC report was a decision about the organization of the content, which almost guarantees that it has little effect on the swings. Every time it produces a major 'assessment report', the IPCC divides its work into three parts, released several months apart. The first report is on the science of climate change, the second is on impacts and adaptation, and the third is on reducing emissions. Predictably, the first to be published gets most of the interest and coverage, while the second and third struggle for attention as, by the time they are launched, they seem like old news. While people close to the subject recognize that each of the reports cover different topics, to outsiders – including news editors – the later reports just seem like another UN publication on climate change.

The result is that media coverage of the reports focuses on the first to be published. This is the report that includes discussion of the scientific evidence that climate change is real. Even though the 2013 report emphasized the high level of scientific confidence that human-caused climate change is real, the fact that it even talked about the topic gave the media license to suggest there is still doubt. In Lakoff's terms, the report allowed the media to activate the frame of climate change as being a matter of scientific dispute, rather than the frame of climate change being an immediate problem that demands urgent attention.

This notwithstanding, you might think at least the reports draw attention to the problems that dangerous climate change would cause, as this – along with the scientific confidence – is one of the topics of the first report. That is true, but the reports fail to frame the problem as one that will directly affect people who hear about it.

The first instalment outlined, at a broad level, how the climate is likely to change in different regions, without giving details about what this is likely to mean in practice. It wasn't until the second instalment, six months later, that readers could find out what the specific consequences of this changing climate would be for them, for example with increased flooding, heat waves and wildfires.

So when the first report was launched and hogged the attention, journalists couldn't report on the likely impacts of a changing climate for their audience. As a result, most of the coverage focused on the traditional story of disputes about the scientific confidence in the findings.

A defender of the IPCC might argue that its primary audiences are governments and policymakers – according to its own documents, the wider public are only among its broader audiences and, in fact, are bottom of the list.[26] But it is hard to take this seriously. If the IPCC issued its reports privately to governments, with no website or press conference, the argument could perhaps be sustained (though it would be a bad approach). But the launch of the assessment reports is a major news event that attracts international coverage. It is one of the best opportunities to challenge climate apathy by framing global warming as a problem that will affect even countries where many people are still apathetic about it. At the moment, the opportunity is wasted. The next report, due in 2021, offers a chance to correct this with a new structure that helps journalists show their audience how uncontrolled climate change would affect them.

Is that all?

So far, we have seen how climate change is often framed as a threat to distant places and animals, and one about which there is scientific dispute. The remainder of the chapter looks at another of the ways in which descriptions of the problem foster apathy – some of the everyday terms used to describe it make it sound, to anyone who isn't already an expert, like it won't be a major problem.

On the last day of 2013, headlines around the world reported the warning of a new study: the planet was on course to warm by 4°C/7°F by the end of the century.[27] I was in England when I heard the report and it was cold, wet and grey. I couldn't help but think that a few degrees of extra warmth sounded quite nice, offering to take an edge off the winter's bleakness. On reflection I knew I was missing the point and that this much global warming would bring disaster – it is close to the catastrophic scenario outlined in Chapter 4. But my reaction reflected another aspect of how the brain works, which we need to understand if we are to beat climate apathy.

The mind interprets the world through two very different processors. One processor is constantly making snap judgements that we are barely aware of. To do this so quickly it uses all kinds of shortcuts, which are mostly helpful but can sometimes lead us astray, including interpreting information through the frames that George Lakoff describes. The other processor is slower, making decisions after more careful judgements. It is the processor we are aware of; it is what we think of as ourselves. Nobel Prize-winning psychologist Daniel Kahneman, who refers to these processors as System 1 and System 2, emphasizes that our unconscious System 1 does much of the work in shaping how we see the world.[28] System 2 won't engage with a problem unless it has to do so, leaving most of the work to System 1. Since System 1 is both border guard and interpreter for the vast amount of information we encounter in the world, it is implicated in climate apathy.

I wasn't alone on that December day in thinking that a few degrees of warming didn't sound too bad. A poll by the website Carbon Brief asked the UK public how many degrees of global warming they would consider to be dangerous.[29] Two in five didn't feel able to guess, while the remainder thought, on average, that climate change would not be dangerous until global temperatures rose by 8°C/14°F.[30]

Most studies of the consequences of global warming don't go beyond temperature increases of 6°C/11°F. At that level of warming, the super hurricanes, city-swamping floods, and droughts that last for generations could be expected to be even worse than those described in the worst scenario in Chapter 4. The last time the world warmed that much, 251 million years ago,[31] 95 per cent of marine animals and 75 per cent of land animals became extinct.[32] In fact, 8°C/14°F is such an extreme amount of warming that there is little research on what the consequences might be. Yet it is the lowest temperature increase at which people think, on average, climate change would start to be dangerous. To put it another way, most people don't know what level of warming would be dangerous, and when asked to guess they choose a number that is far larger than they would choose if they knew more about the subject.

This isn't a lament about the state of education today or the laziness of the public for not paying enough attention to such an important issue. The problem is in how we talk about climate change.

When we refer to global degrees of warming we are talking in a code that has two layers of encryption. The first encryption is the use of global temperatures. No-one experiences global temperatures – we all experience temperatures where we are at any time – and it is hard to develop an emotional attachment to a world temperature. Even if you are a citizen of the world who is worried about the impact of climate change on distant places, global temperature changes don't tell you much about this either. An average global temperature change doesn't mean the same increase everywhere – some places will warm faster than others.

The second encryption, the use of temperature change, is even more of a problem. The numbers used are always small – a few degrees – and so have little emotional impact. For anyone who hasn't thought about climate change enough to have an immediate reaction to a description of climate change with small numbers, a warning that the world will warm by a few degrees is generally filtered out by System 1 before System 2 is engaged. To realize it is important, the person hearing about the warming has to over-ride their System 1 – which they can only do if they already understand the significance of those small numbers.

The phrase '6°C/11°F of warming' is hopelessly ill-equipped to convey the horror it signifies. With this much global temperature rise, we would be likely to see much of Australia baking, fires raging across Canada, the UK drying up and the center of the US turning into a permanent Dust Bowl. Floods and super-storms would ravage coasts across the planet and huge areas would disappear to the sea. As the world heats, we would run an ever-greater chance of triggering tipping points where melting glaciers or permafrost, or the collapse of rainforests, could unleash a burst of extra heating that would make everything even worse. None of this is conjured up by bland phrases like 'six degrees' and, while the person saying it might understand its significance, anyone who isn't already knowledgeable about the topic could easily miss the point.

This isn't going to be fixed with just a bit of public education. Global average temperature change is an unhelpful concept even for people who know quite a lot about climate change. Whenever anyone – apart from the few who are so knowledgeable about climate change they respond emotionally to the small numbers in question – hears a term like 'six degrees' they have to activate their System 2 and refer to an internal codebook to translate the term into something that tells them what it means. And even then, there's yet another problem – most of our codebooks aren't detailed enough to help with the level of nuance that is needed. I have spent much more time than average reading and talking about climate change but, until recently, my codebook still looked something like this:

> 2°C/4°F: bad, but survivable if you are fairly well off and don't live by the sea or in a place that is already hot and short of water.
>
> 4°C/7°F: somewhere between 2°C/4°F and 6°C/11°F.
>
> 6°C/11°F: awful. Much of the world will become effectively uninhabitable.

This leaves obvious gaps. Suppose I see a projection of 3°C/5°F of warming. My codebook is of little help here – all it does is suggest the result would be a bit worse than 2°C/4°F but quite a bit less bad than 6°C/11°F. I have a similar problem when I hear discussions about limiting warming to 1.5°C/3°F, as world leaders agreed in principle at the Paris climate conference. I know it is better than 2°C/4°F but I would have to look up the answer if I was asked what it would actually mean in practice. We can't seriously expect that most people will become so knowledgeable about climate change that they will carry these codes in their heads – the problem is in the use of the numbers as if they convey much information.

What is more, the codebook is global. There is nothing in it to tell me what, specifically, these changes would mean for me, my family or the places I know. My scale of bad to awful doesn't tell me, for example, how often I should expect my city to flood or how many days I should expect that are dangerously hot, if average world temperatures rise by any particular number of degrees. If a series of small numbers is what most people hear about climate change, it is no wonder so many think it won't be dangerous until temperatures rise by 8°C/14°F.

This isn't the only example of a term that doesn't work at an emotional level yet is widely used in public discussions of climate change. There is a similar problem in discussions of how high and how quickly sea levels might rise. It is one of the widest-known consequences of climate change, but the way it is described often makes it seem unthreatening for non-specialists.

In the IPCC's worst-case scenario of high emissions, sea levels at the end of the century are projected to rise by between 0.8 and 1.6 centimeters a year.[33] To my System 1, that doesn't sound like much. The city where I live, London, is one of those sometimes talked about as being under threat as its main river is tidal. But my front door is about 13 meters above the level of the river, so with a rise of one centimeter per year, it sounds like I could live there for 1,300 years before I would need to start buying sandbags.

Again, my System 2 slowly wakes up and tells me that this is not the point. The flooding of everything around the river should worry me long before the waters reach me. And, more importantly, average sea-level rise isn't the only problem. What also matters is what happens at the extremes – what changes in average sea levels would mean during high tides and storm surges. By focusing on averages and annual increases, we provide false reassurance for System 1, which means we are counting on our audience employing their System 2 to realize the significance of what they are hearing.

This would not be such a problem if small-number descriptions of global average temperature change and sea-level rise were used only in scientific and policy discussions. Plenty of policy debates are held among the small number of people who are experts in the field and it is normal for those debates to use technical concepts that non-specialists are unfamiliar with. I'm not bothered that I can't tell you the level of pesticides that can be safely left on an apple, because I'm confident that the question will be tackled by people who do know the answer.

But climate change is different. Discussions about it aren't confined to technical forums in the way debates about pesticide levels are, and responses to it can't be left to just a few policy specialists. Yet technical terms about climate change are used in mainstream discussions as if they are widely understood.

This reliance on abstract concepts like average global temperature change and annual sea-level rise – described with small numbers – may have contributed to the problem of many people misunderstanding the likely consequences of climate change. One study found that most people in the UK largely, and correctly, think that climate change will be likely to lead to more intense rain and rising sea levels, both causing flooding,[34] but that they are much less likely to think that it will cause heat waves.[35] What's more, when asked how worried they would be if the country did experience heat waves, more said they wouldn't be concerned than said they would be. This isn't particularly unexpected – heat waves in the UK are still relatively rare and are usually treated as a pleasant surprise, even though, as we saw in Chapter 4, extreme heat waves can be deadly for old and other vulnerable people. Unexpected or not, it suggests that most people aren't familiar with the likely impacts of climate change.

That research also hinted at how misunderstandings like this can be resolved. Buried in the data is a strange result that, at first, seems to be wrong. While only 33 per cent think that heat waves will become more common by 2050, 52 per cent say they expect more old people's health to suffer *as a result of heat waves* by then.[36] How can more people think that heat waves will harm old people's health than think that there will be heat waves at all? We don't see the same with questions on potential positive effects of climate change – for example, roughly the same proportion of people think that we will see warmer winters as say that we will see warmer winters that lead to fewer people dying from the cold.[37]

This isn't a mistake in the research. Instead, it is an example of an error that System 1 sometimes makes, known as the conjunction fallacy. Its mistake is to guess that a familiar event, described in detail, is more likely than a generic event, even though the familiar event is a specific example of the generic event. For example, only 20 per cent might agree with the statement 'a war between the US and China will break out in the next decade', but 40 per cent might agree with the statement 'in the next decade, tensions over Taiwan will escalate, drawing in international powers until the US and China are at war'. This is about the power of familiar stories. People in the UK are familiar with the story of old people's health suffering in extreme weather, but they aren't familiar with the story of heat waves, so they find the specific story more believable than the generic one. Examples like this reflect the importance of framing climate change in ways that convey its consequences to non-specialists.

It is also interesting that there is such a difference in responses to the possible positive effects of global warming compared with its negative effects. People seem more likely to associate climate change with causing harm, such as people dying from heat waves, than doing good, such as people not dying because of warmer winters. This might seem encouraging, as it suggests that climate change is already better understood as a threat, rather than as a benefit. But it is not entirely good news for tackling apathy. While some research suggests that climate change may not in fact reduce winter mortality rates in colder countries,[38] in the short term it may still bring some benefits to areas at higher latitudes, such as better growing conditions for food crops.[39] It would be dishonest for those of us worried about global warming not to mention this. And, if honesty isn't reason enough, it is bad campaigning to ignore arguments that appear to undermine your position. It is far better to acknowledge and contextualize apparently challenging facts – 'climate change will bring some benefits but, even in richer countries, these will be greatly outweighed by the damage it causes' – than to let critics be the only ones making those arguments and overstating their significance. Otherwise, people will doubt the honesty of the other parts of our argument when we talk about why climate change is a serious problem that needs urgent attention.

Much of how climate change is described makes it seem unthreatening for non-specialists, particularly those who are apathetic. This is partly because people talking about climate change often focus on global and distant threats, including those to animals and far-off places. Such warnings may help to mobilize people who are already worried about the problem, but they do little for the swings. The debate also constantly draws on technical terms that have little emotional impact, ignoring lessons from psychology about how the brain filters information. Together, these common ways of describing climate change reinforce a framing that does little to challenge apathy. Chapter 8 looks at how we can talk about climate change in ways that address these problems and make it seem more important to the swings. But first, we will see one other way that climate apathy is fostered – the presentation of global warming as a battle between Left and Right.

7

Do you have to be leftwing to worry about climate change?

Climate change is widely seen as an interest of leftwing environmentalists, and some campaigners do little to resist this label. But many people who are apathetic about climate change don't identify with the Left and so assume that the threat is exaggerated for political reasons.

'When climate-change deniers claim that global warming is a plot to redistribute wealth, it's not (only) because they are paranoid. It's also because they are paying attention.'
– Naomi Klein, author and social activist[1]

As the head of the Organisation for Economic Co-operation and Development (the OECD), Ángel Gurría is no anti-capitalist. So it made a minor news story when he told a development conference that fossil fuels – the basis for two centuries of economic growth – are 'the enemy'.[2] It seemed surprising to hear a call for radical action to cut emissions from the leader of an organization that describes itself as 'a steering group for the world economy'.[3]

But it shouldn't have come as a surprise. For years, people who aren't members of any leftwing conspiracy have argued for significant measures to cut emissions. Among them is the former British Prime Minister, Margaret Thatcher, who told the UN in 1989: '[It] is mankind [sic] and his activities that are changing the environment of our planet in damaging and dangerous ways.'[4] Others in the crowd of unlikely radicals include the militaries of

Australia, Canada, the UK and the US,[5] NATO,[6] the International Monetary Fund,[7] Goldman Sachs[8] and Coca-Cola.[9] Indeed, a 2014 study found that rightwing governments are just as likely to pass climate laws as leftwing ones.[10]

Given this vocal support and material action from across the political spectrum, Gurría's denunciation of fossil fuels was nothing out of the ordinary. Yet, for some reason – however many global businesses, conservative politicians and members of the military-industrial complex warn about climate change – it still seems like an issue that 'belongs' to the political Left. When conservatives talk about climate change, it is hard to shake the feeling they have stepped outside their natural territory.

To test whether this feeling is widespread, I surveyed public opinion in the UK and US. The poll, conducted by PSB, found that climate change really is seen as a leftwing issue. In the US, 46 per cent said liberals are the people who are most worried about climate change, compared with only 8 per cent saying conservatives are most worried. This made it the most liberal-leaning issue of the eight we tested – more than both inequality and housing. The polarization was less in the UK, where only 22 per cent said they think leftwing people are the most worried about climate change, although that was still more than four times as many as said they thought that rightwing people were generally most worried about it.[11] So both the UK and US public seems to associate climate change with the political Left.

This chapter looks at why this is and why it matters. It is sometimes argued that only leftwing approaches can prevent dangerous global warming, suggesting that the link between climate and the Left is inevitable. Some advocates of action on climate change also suggest that the threat it poses represents an opportunity to revolutionize economies in ways that they have long advocated. The chapter examines how these arguments – and the association of global warming with the Left – influence climate apathy.

Left or Right, open or closed

Before going on, we need to talk about this term 'leftwing'. It may seem well defined, but when it comes to the debate about climate change, the term can be misleading.

One common way of defining the Left is to characterize it as the belief that governments should address ills that the free market or other non-state actors can't, or won't, fix[12] (in some countries this is usually described as 'liberal', but in others 'liberal' refers to right-of-center politicians, so I am using the term 'leftwing' to avoid confusion). Preferences for these interventions are more common among people on the Left than they are among those on the Right. Towards the right end of the left-right scale are neoliberals, who believe the state's role should be limited to protecting property rights, free markets and free trade.[13] At the left end are socialists, who believe in government planning of the economy.[14] Traditional conservatives, to the right of center, are cautious about state intervention, but are principally concerned about maintaining order, and are willing to accept some state intervention to relieve pressure for radical change that might otherwise accumulate and eventually lead to more chaotic change.[15] Modern politics in most richer countries is dominated by people who are at neither extreme – despite the effect of leaders like Reagan and Thatcher on shifting politics to the Right, most mainstream politicians still believe markets should be regulated, for example in healthcare, consumer products and where they affect the environment. Not everyone who supports any such intervention is considered to be leftwing. Nevertheless, a preference for interventions to address the unhappy consequences of free markets, bad luck or bad decisions is more common on the Left than it is on the Right.[16]

But even with this definition of the Left, there is still a problem with the attempt to divide the world along a Left-Right axis. That division ignores a separate schism that, while often overlooked, is in fact essential to how those who worry about climate change are seen. Most major rich countries now have an anti-immigrant political movement. In some cases these are distinct political parties, such as the French Front National and the UK Independence Party (UKIP); in others they exist within major parties, such as that which elected Donald Trump. While these groups are often described as rightwing, the reality is often more complicated. Many supporters of UKIP, for example, think that wealth is unfairly distributed, that big businesses are run for the benefit of their owners at the expense of workers, and that the government should narrow the gap between rich and poor,[17] all traditionally leftwing positions. But, while many UKIP supporters may have those views in common with many leftwing environmentalists, the two groups profoundly disagree on questions of immigration, criminal justice, foreign aid and climate change.[18] This split is sometimes labelled 'open against closed' – openness to change against resistance to it, or enthusiasm for the future against nostalgia for the past. This applies to views on social issues such as immigration and sexuality, and to economic issues such as free trade and other matters linked to globalization.[19] Nationalism is often a form of closed politics and both the UK's vote to leave the European Union and the US presidential election victory of Donald Trump reflect this division, with electoral successes for closed arguments.

Combining these two axes gives us four quadrants: left-open, left-closed, right-open and right-closed. This map of where major

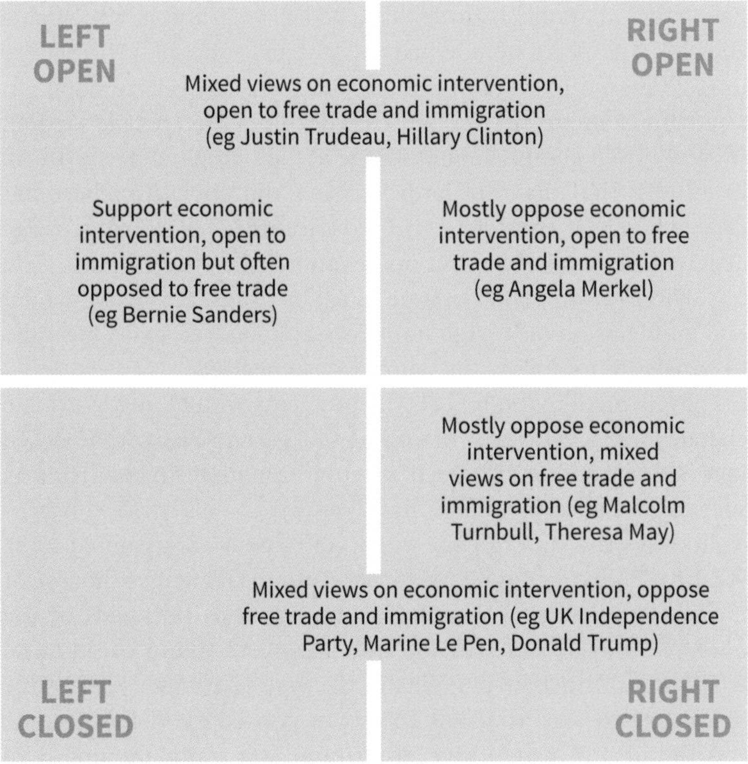

LEFT OPEN

RIGHT OPEN

Mixed views on economic intervention, open to free trade and immigration (eg Justin Trudeau, Hillary Clinton)

Support economic intervention, open to immigration but often opposed to free trade (eg Bernie Sanders)

Mostly oppose economic intervention, open to free trade and immigration (eg Angela Merkel)

Mostly oppose economic intervention, mixed views on free trade and immigration (eg Malcolm Turnbull, Theresa May)

Mixed views on economic intervention, oppose free trade and immigration (eg UK Independence Party, Marine Le Pen, Donald Trump)

LEFT CLOSED

RIGHT CLOSED

figures and parties, with clearly distinct views, fit into these quadrants shows the benefit of using both axes.

Any placement in the grid can be disputed, including where the center is. Placing politicians from different countries on the same scales can also be misleading, implying more similarities between political cultures than is really accurate. But even though this model is imperfect, it reveals a point that is crucial for thinking about how climate change is perceived. Rather than leftwing-ness, the consistent characteristic of people who are worried about climate change is openness. There are people who sometimes express leftwing views but are closed and suspicious of issues like climate change – like some nationalists. Likewise, there are people who are rightwing economically but open and worried about climate change, like some conservatives such as Angela Merkel.

The open-closed scale is mentioned far less often than the Left-Right one, and people worried about climate change are, as we have seen, generally considered to be leftwing. But when climate change activists are stereotyped, it is not just as people who believe in government intervention to provide a strong welfare state, but also with elements of the open-closed scale. When people say that those worried about climate change are leftwing, what they have in mind is that they are both leftwing and open. This becomes clear when you consider that the caricature of a climate-change activist is someone who not only wants higher taxes on the rich, but who is also liberal on issues such as drugs, sexuality, gender and immigration (although not necessarily on free trade, attitudes to which do not follow this caricature so neatly).

This distinction also reflects the problems with allowing climate change to be seen as an 'environmental' issue. Some on the closed Left are suspicious of aspects of environmentalism, particularly species conservation. They have tended to see these concerns as luxuries that interest the open Left – who are assumed to be wealthy and cosmopolitan and are often labelled as liberal or metropolitan elites – who fail to understand the needs of the poorest. They worry that too much concern about the natural world is a distraction that gets in the way of human well-being. The desire to save the spotted owl or crested newt has become shorthand for unworldly environmental campaigners who care more about obscure animals than about humans. This was exacerbated by the 2008 economic crisis. Many center-left political parties have prioritized demonstrating their economic competence over tackling challenges like climate change.[20]

The term 'leftwing' is clearly an inadequate description for those worried about climate change. There are people who are leftwing who don't think it is a serious problem, and there are people who are rightwing who do. Nevertheless, the label is commonly used to talk about those who worry about it. So, while the term 'open left' would be more accurate, the rest of the book continues to use the term 'leftwing' – but does so with the specific meaning of people who are generally both in favor of government interventions to address failures of the free market (the Left) and who are socially liberal, comfortable with the modern world and have little yearning to return to a more traditional age (the open).

System change, not climate change

If, as we saw at the start of the chapter, many rightwingers publicly warn of the terrors of climate change, why should anyone think it is a leftwing concern? It is not self-evident that worries about climate change should be associated with the Left. Conservation has previously been associated with preservation of hunting and recreation grounds for the wealthy, at least as much as with protecting species and wild places for their own sake. Rachel Carson's 1962 book, *Silent Spring*, has been credited with awakening the modern environmental movement, yet her argument that pesticides were accumulating in the environment wasn't inevitably leftwing.[21] Her findings indicated that people should be restrained in their use of such chemicals, a warning that might be considered leftish in that it would limit the activities of the private sector, but just as easily fits with the conservative creed of accepting some restrictions to protect the world as it is and to avoid later radical change.

In part, the labelling of climate change as a leftwing concern reflects how it has been defined by its opponents. The book and film *Merchants of Doubt* show how environmentalists were presented as the enemy by the US Right in the early 1990s, when the collapse of the Soviet Union left them bereft of a superpower opponent against which they could position themselves. Seeking to fill the gap left by the communists, the Right argued that environmentalists were anti-capitalist leftwingers who threatened the American way of life, and so suggested the enemy within should take the place of the enemy without.[22]

But climate change's leftwing label isn't a complete fiction – it really does interest people on the Left more than those on the Right. In the poll mentioned earlier – that found climate change is generally seen as an issue that worries leftwing people – I also asked people how worried they were about each issue. Those in the US who had identified themselves as liberal were nearly five times as likely as conservatives to say they are very worried about climate change: 54 per cent compared with 11 per cent. In the UK, those who said they were leftwing were more than three times as likely as those on the Right to be worried about it: 38 per cent to 11 per cent.

In fact, not only is the leftwing label not a fiction, but this perception might be partly the result of the efforts of those who sought to paint climate change as a leftwing concern. The polarization is a relatively new phenomenon. Research in the US shows that, until 1991, self-identified Republicans and Democrats were just as likely to say they were environmentalists, but by 2016 Democrats were twice as likely to do so.[23] The label has become much less popular, with 78 per cent identifying with it in 1991, compared with 42 per cent in 2016. By presenting environmental protection as a leftwing priority, rightwing opponents of environmental protections made the subject a question of identity. For those on the Right, it became a label to identify against, and – even though the polarization was invented by those on the Right – environmentalism became a label for those on the Left to identify with.

This isn't the full picture, though. Climate change isn't perceived as a leftwing issue only because it worries more people on the Left or because of how it is presented by the Right. It is also seen as being leftwing because of some of the arguments that environmentalists use when they talk about addressing the problem.

Two alternative approaches to dealing with climate change divide those who argue for action on it. Some people believe that the only appropriate response to the threat of climate change is to overthrow the growth-focused model that dominates world economies. These people could be described as 'revolutionaries', since they argue that it is necessary to change fundamentally economic (and therefore political) paradigms. But others, who agree with the first group about the severity of the threat of global warming, believe that capitalist markets, with direction from governments, can tackle the problem. People with this view might be called 'incrementalists', as they tend to believe that the best hope of avoiding disaster is through a huge number of changes that are, individually, not revolutionary, even if the cumulative effect would be transformational.

Not only do these two sides offer sharply different visions of how dangerous climate change can be prevented, but the differences have implications for how the swings perceive the politics of the problem.

Many – although not all – of the revolutionaries argue that the world can never prevent dangerous climate change as long as it sticks with an economic model based on growth. According to this argument, set out in detail by the economist Tim Jackson in *Prosperity Without Growth*,[24] although the world can reduce the emissions produced by each unit of economic activity, it can't do so more quickly than global economies grow. That would mean that the volume of emissions produced each year would continue to increase, or at least would not fall much (and remember that annual emissions must fall dramatically for the world to avoid dangerous global warming). Therefore, the only way to avert dangerous, or catastrophic, climate change would be for some economies – presumably those that are already the most developed – to stop expanding, or even to shrink to allow others to grow.

It is inescapable that this sounds leftwing. Stopping economic growth would only be possible with extensive intervention. Not only would some measure be needed to prevent growth, but there would almost certainly have to be massive wealth redistribution to compensate those who had previously counted on the hope of future growth to alleviate their poverty. Other measures would also be needed to stop the finite amount of wealth accumulating into fewer hands. These redistributive measures sound familiar to some on the Right who consider climate change a leftwing plot.

But, while revolutionaries may acknowledge that this would be disruptive, they often argue that it is the only way of avoiding climate disaster, and that the absence of growth is a vastly lesser evil than this warming.

But plenty of people who are worried about climate change disagree with the revolutionaries. They argue that, with the right approach, the emissions produced by each unit of economic activity can be reduced more quickly than economies continue to grow, meaning that annual emissions would fall. This may not cut emissions as quickly as abandoning growth would, but, they argue, it is exceptionally unlikely that most of the public will agree to abandon growth-focused capitalism. Since calls for an end to growth-focused capitalism won't be taken seriously by people who aren't already sympathetic, such a focus may do nothing to limit global warming. If, instead, people who are worried about climate change prioritized incremental change, they might achieve much more.

Of course, the planet's climate doesn't care about public opinion – of far more importance is whether it really is possible, in a growth-focused system, to cut emissions enough to avoid dangerous warming. So while this chapter focuses on what the perception of climate change as leftwing means for how the issue is viewed, Chapter 9 returns to this debate to discuss what it means for efforts to cut emissions.

As well as the view that economic growth is irreconcilable with a stable climate, there is another kind of revolutionary argument that may contribute to the perception of climate change as leftwing. So far, I have described the view of revolutionaries as: 'Economic growth will inevitably lead to catastrophic climate change, so, like it or not, we have no choice but to revolutionize the economy.' But that doesn't describe every revolutionary. Instead of arguing that there is no alternative but to respond to climate change by transforming the focus of global economies, some instead say that climate change provides a welcome opportunity to transform economies, even if that isn't strictly necessary for cutting emissions.

This argument has recently been given prominence by the Canadian writer, Naomi Klein. Her best-selling 2014 book, *This Changes Everything*, says that global warming represents an opportunity to do things she already supported: 'Climate change is inconvenient only if we are satisfied with the status quo, except for the small matter of warming temperatures. If, however, we see the need for transformation quite apart from those warming temperatures, then the fact that our current road is headed toward a cliff is, in an odd way, convenient.'[25] In fact, she rejects the view that economic transformation is the only way of avoiding dangerous climate change: 'It would be reckless to claim that the only solution to this crisis is to revolutionize our economy and revamp our worldview from the bottom up.'[26] Instead, she says, 'Climate change is our chance to right... festering wrongs.'[27]

This argument – that climate change is an opportunity to advance other goals that were desirable anyway – makes it easy for opponents of measures to cut emissions to claim that the risks must be exaggerated. In Chapter 1 we saw how 'motivated reasoning' can lead defenders of unregulated free markets to convince themselves that climate change isn't a serious threat. But if someone hears an argument that climate change requires the world to make radical changes, and knows that the person making the argument had wanted those changes in the absence of global warming, they too might wonder whether motivated reasoning is a factor and that revolutionaries may be exaggerating the threat.

Not every revolutionary shares the view that radical economic change isn't essential for controlling warming but is desirable for other reasons. Many people believe it is both necessary and desirable. Some have no position on whether economic revolution is desirable in its own right, but are convinced it is essential for limiting warming. There may be people who wouldn't have supported economic revolution if it wasn't for climate change, but regretfully think it is the only way to avoid disaster. And not every revolutionary seeks the same type of economic transformation – some focus on ending economic growth, while others prioritize transforming how resources are owned. Nevertheless, as this chapter goes on to consider the consequences of climate change being seen as a leftwing issue, we should be conscious that some revolutionaries openly argue that the threat is an opportunity to advance goals that seem leftwing.

Even so, just because some people use the threat of climate change to argue for economic revolution, it still isn't inevitable that the issue itself should be perceived as an interest of the Left. Not everyone who is worried about climate change does so. And we have already seen that there are other people, who aren't leftwing and who are worried about the climate, yet the problem doesn't seem to be associated with their views.

One explanation for this might be that, from the perspective of most people who don't pay close attention to climate change, the debate seems to be dominated by environmental campaigning organizations. It probably doesn't seem that way to someone who closely follows the debate and listens to the detailed discussions involving a wide range of people and groups about how the world can limit climate change (the focus of most international conferences and specialist websites). But my focus here is on the small amount of coverage of climate change in the most-seen general media outlets. Most of this coverage isn't about what the world can do to deal with the problem – the focus is principally on supposed disagreements about whether it is happening, with occasional coverage of protests. These aspects of the debate are dominated by campaigning organizations, while other groups tend to avoid joining in.

An example of this was during the 2015 Paris climate conference, one of the rare times when climate change was a top news story. During the second week, there were nearly as many stories in UK newspapers about climate change as there were about immigration, health, welfare and crime put together.[29] And during that week, environmental groups got much of the coverage. While slightly more than half of the stories focused on governments, the next largest group to be featured was environmental NGOs. In comparison, businesses and industry bodies received far less attention.[30]

This isn't to suggest revolutionaries get more coverage than incrementalists in the most prominent outlets – they almost certainly don't. Many of the most prominent environmental groups are closer to being incrementalist than revolutionary – in fact, some groups are often criticized for this – but that distinction is lost on most people. What most see is simply that an environmental group is speaking, and, as we have seen, the concept of environmentalism is polarizing. While 56 per cent of

Democrats in the US consider themselves to be environmentalists, only 27 per cent of Republicans do.[31] Since the second swing group, the *Cautious*, are as likely to be Republicans as they are to be Democrats, this suggests that associating global warming with environmentalism makes it difficult to persuade more people that it matters. To many people, environmental groups are voices of the Left, yet they are widely heard as spokespeople for climate change.

Environmental groups can't be faulted for getting media coverage – talking about climate change is what they are supposed to do. But many people's interest in the issue has waned in recent years. For many, the financial crisis from 2008 was a turning point. In their eyes, climate change had until then appeared to be one of the most pressing challenges facing the world. Particularly for moderately open politicians of the Center and Right, being seen to tackle climate change was previously a sign of modernity,[32] but as the crisis developed, many of them increasingly saw global warming as a lower priority than reducing public debt.

Climate change doesn't get much media attention, but it isn't invisible. Yet, when it is covered, environmental groups – including revolutionaries and incrementalists – are more widely heard than others who share their view that climate change is a major threat, and so it is these groups that are seen by many in the wider public as typical of those who are worried about the climate.

Left alone

The leftwing reputation of worries about global warming worsens climate apathy, in two different ways. The first, about motives, is visible; the second, about identity, is more subtle.

More than at any other time in generations, governments today face suspicion about their motives, which makes it harder for them to persuade the public that it is right to take significant action on any issue. Most people now doubt that national politicians are looking out for the good of the country, rather than for their own careers.[32] This spells trouble for the measures that would be needed to avoid dangerous warming. Meaningful action would depend on major changes to taxes and public spending – whether through incentives for clean energy or taxes on polluting activities – which are harder than ever to bring about when the public's assumption is that politicians are mostly liars. This is made even worse by climate change being seen to belong to one corner of the political scales.

But the problem goes beyond distrust of politicians – there is also a receptive audience for questions about the motives of the people who call for radical changes to cut emissions. Some opponents of action suggest that those pushing for emission cuts are doing so for reasons other than wanting to avert extreme warming, and that the costs of the actions they propose would outweigh the benefits. In richer countries with high emissions, critics often claim that the costs imposed on poorer people to subsidize renewable electricity aren't worth the benefits those people will get from reduced climate change. And at the international level, it is argued that expecting the world's poorest countries to limit their emissions does more harm than good to their citizens. At the heart of this argument is the suspicion that people calling for action on climate change care more about abstract environmental values than they do about human well-being, and that they are prepared to make people worse off if that is what it takes to protect the environment. This works well as an accusation, as it builds on caricatures about the open Left to create a neat storyline with an identifiable villain.

All of these accusations can be refuted. A consequence of the connection between leftwing values and climate change is that most people who worry about the climate are also worried about poverty. Pretty much every serious organization or campaigner who tackles questions about renewable subsidies or energy access in poor countries goes to great lengths to make sure their proposals do the most good for poorest people. This is why, for example, most environmental organizations are now doubtful about the overall benefits of biofuels that might compete for land with food crops. They also increasingly argue that clean sources of power will be better for many of the poorest people in developing countries than dirty sources like coal. The question of whether it is worth tackling climate change at the possible cost of slower growth in the short term, is still an argument among economists – but the weight of evidence seems, by far, to be on the side of those who conclude that the long-term economic benefits of acting now to reduce warming would easily outweigh the benefits from prioritizing economic growth.[33]

But the aim of attacks like these isn't to win the argument – it is to create doubt about the motives behind those efforts and, by implication, about whether significant measures to cut emissions really are needed. For the attacks to succeed, the people hearing them don't need to be completely persuaded, they just need to develop enough doubts to think the truth might be somewhere between the two sides and that it is worth holding off with stronger climate policies for a bit longer. The success of this argument is reflected in a UK poll, which found that the public believe, by a margin of two to one, that the government uses climate change as an excuse to put up taxes.[34] Climate apathy is a direct consequence of such views.

There is also a second way in which the association of climate change with the Left fosters apathy. This process is more subtle – it is based on political identity and the lack of it.

For people who consider themselves to be rightwing or conservative – about 30 to 40 per cent of the population[35] – and who think climate change is an issue that belongs to the Left, psychology predicts a problem. As we saw in Chapter 5, humans have a tendency to split the world into ingroups and outgroups – people they identify with and those they identify against. People tend to assume that others in their ingroup are correct and those in their outgroup are wrong, and they often interpret any evidence they are presented with in ways that reinforce that assumption. Since climate change has become associated with a clearly defined group (leftwing open people), people who see the open Left as an outgroup are much less likely to take climate change seriously. They have to fight against their predisposition that, as an issue of interest to people they see themselves as opposed to, climate change can't be that important.

And that is just the one in four people who think of themselves as rightwing or conservative. More than 40 per cent of the public don't have strong political allegiances, thinking of themselves either as politically centrist or just not particularly interested in politics.[36] Many of those in the middle believe that climate-change partisans of each side tend to exaggerate. They perceive both groups to be arguing from entrenched positions, seeking to score points and win childish victories rather than trying to make the world better. If someone sees the two sides as being as bad as each other, it is not surprising that they would conclude that the 'true' answer to climate change must be somewhere in the middle.

This is reflected in the segmentation research. The US study shows how sharply polarized climate change is between political wings. Among those in the base segment (the *Alarmed*) nearly half consider themselves liberal – more than three times as many as say they are conservative. As the ingroup-outgroup model predicts, the people who are least worried about climate change – the two critic segments – have nearly opposite political views, with 61 per cent and 76 per cent respectively thinking of themselves as conservatives, and just 6 per cent and 3 per cent saying they are liberal. Those in the swing groups, the *Concerned* and the *Cautious*, have views in the middle. In nearly all cases, more swings describe themselves as moderates than as either conservative or liberal. So, while rightwing people's view of the Left as an outgroup means that they are particularly resistant to action on climate change, which they think is supported by the Left, swing groups' perception of climate change as a battle between Left and Right means they are more likely to be apathetic about it. This also helps to explain the rarity of everyday conversations about climate change, as we saw in Chapter 5. Most people have a feeling that it is inappropriate to talk about controversial topics like politics and religion with those they don't know well. The framing of climate change as a polarized battleground means it is seen as a controversial topic that is best avoided in polite company.

What is more, climate change suffers from a particular problem as a result of its association with the open Left. The puerile, offensive and often brilliant TV cartoon *South Park* aims to ridicule everyone, and this group has been an easy target. In a 2006 episode, 'Smug Alert!' (the title gives a clue about where this is going) one of the characters buys a hybrid car to show off his concern about the planet.[37] When his neighbors are insufficiently impressed and he decides they aren't sophisticated enough for him, he moves to San Francisco, a city whose residents are portrayed as so self-satisfied they enjoy smelling their own farts.

Of course this is a parody – but it works because enough people see it as having an element of truth. The danger in someone declaring that other people's lifestyles are destroying the planet and that they personally are living more ethically is that many others will perceive them to be declaring themselves superior. This is unlikely to persuade others to change their lifestyle. If you are aware of this stereotype and you are worried about climate change but don't know whether the person you are talking to feels the same, you may be worried that raising the subject could make them think of you as a bit, well, smug. In this way, the identification of climate change with environmentalists of the open Left means not only that the motives of people worried about it are questioned, but also that those who don't identify with that group are inclined to dismiss it as the interest of a small, and slightly odd, minority.

We have now reached the end of this section, which has explored the causes of climate apathy. It started by looking at how, and how often, climate change is discussed, both in the media and in everyday conversation. We saw that apathy is partly caused by how the nature of climate change itself interacts with aspects of human psychology to make it easier for many people to ignore an apparently slow-moving and distant threat. But, as this chapter and the previous one have shown, there are some factors behind climate apathy that are caused by how those of us who are worried about global warming often talk about the problem. We unintentionally make it less interesting to the people whose attention and interest we need to make up a climate majority. Descriptions of the threats that climate change pose often fail to capture the attention of the swings, and the identification of it as an interest of environmentalists on the open Left has made some people hostile towards it and others apathetic.

To adapt Karl Marx:[38] so far we have only interpreted climate apathy; the point is to change it. The final section looks at what we can do to challenge the causes of apathy and to build greater support for the measures that will be needed to avoid disaster.

Part 3
How to beat apathy

8
The pointy end

People who are apathetic about climate change will only begin to pay attention to the problem if they see what extreme warming would mean for the people and places they care most about. They also need to see that it is not too late to avoid disaster.

'"First lesson," Jon said. "Stick them with the pointy end."'
– George RR Martin, *A Game of Thrones*[1]

Over the last three chapters we have seen a host of factors that create climate apathy. Some of those factors are inherent to global warming, such as its complexity and distance, which interact poorly with human psychology, while others are created and avoidable, such as the way the consequences of climate change are described and its association with environmentalists on the open Left.

While a core of about one in five people is deeply worried about climate change and willing to make sacrifices to deal with it, and a similar-sized group are opposed to cutting emissions, a much larger group – slightly more than half the population – are in the middle. In general these people neither disbelieve climate science nor are alarmed about climate change. Most of the time, they just don't think about it.

Their apathy spells trouble for efforts to avoid dangerous global warming. On the positive side, many high-emitting countries have started cutting their emissions, and most people who are apathetic about climate change support the idea of dealing with it. But most people in the swing segments don't really see the point of making major adjustments to their lives for the sake of the climate. If high-emitting countries ask their citizens to make more significant changes for the sake of climate change, they will be unlikely to find a receptive audience. To avoid such a confrontation, governments of those countries have focused on the relatively painless changes. Major decisions, like airport expansion, are made without much debate about their implications for the climate.

This section looks at how we can change this. First, this chapter focuses on increasing attention to climate change. The swings will never be persuaded that they should support measures to cut emissions if they don't see why the problem matters to them. The next chapter looks at how the climate debate can become less partisan, so the issue is recognized as one that everyone should care about, rather than being marginalized as the interest of a particular group. The final chapter focuses on how we can build an optimistic vision about tackling climate change, so it shifts from being a bleak and insurmountable problem that most people prefer to ignore, to being seen as a challenge that the world has the power to overcome. Each chapter describes how climate change could be talked about in ways that could help overcome the causes of apathy, then fleshes out those descriptions, with details about different aspects of the argument. The aim is to provide a set of descriptions of climate change that readers can draw on when they are talking about the problem, particularly with the swings.

Really real climate change

The end of the world is imminent in Terry Pratchett and Neil Gaiman's 1990 book, *Good Omens*, and for various reasons one of the main characters – Adam, an 11-year-old boy who happens to be the antichrist – is dividing up the planet among his friends. Each will get some parts of the Earth, but soon Adam's friends realize he hasn't left any of the continents for himself. Adam explains this is because all he wants is Tadfield, the small English town where they live. But his friends won't let him:

'"You can't," said Wensleydale flatly. "They're not like America and those places. They're really *real*. Anyway, they belong to all of us. They're ours."'[2]

As so often with Pratchett and Gaiman, this apparently off-hand joke delivers a smart insight into human nature. We all know, logically, that distant countries are real places that actually exist. But it is the places we know, live in and grew up in that are, emotionally, the most real to us. What's more, those places usually matter to the people we know, unlike distant places, which mostly matter to strangers. Of course, owning a vast, if distant, continent should be better than owning a small local town, but it can be hard to persuade our emotional selves that this is true. For most of us, the places we know are more important than those we don't. This is crucial for how most people think about climate change.

We saw in Chapter 6 that climate change is unintentionally made to seem less interesting by the way it is often described with warnings that most people don't find particularly threatening. The language and examples that are typically used to describe the threat it poses – including by people worried about it – fail to attract the attention of the many people who aren't already familiar with it. Much of the discussion about its likely consequences focuses on what it means for animals and distant places, and the terms used to explain likely changes often make it seem less threatening than it really is. However much we try to care about those distant places and non-human animals, for most people, they are not really *real*.

The strongest supporters of nationalist arguments like those used by Donald Trump appear to be those who are already opposed to cutting emissions, rather than the people who are apathetic about the problem[3] – but the electoral success of nationalists nevertheless puts them in a position to argue against international co-operation. Chapter 10 discusses how dealing with climate change can make it possible, in time, to undermine this isolationism. But that may not be possible immediately. First, we need to challenge the argument that climate change will overwhelmingly be a problem only for people in other countries.

To get more swings to pay attention in the first place, climate change should instead be described in terms of its impact on the life they know. The swings, who don't reject climate science but still don't think much about climate change, generally have no clear understanding of how it is likely to affect them, their family and their community. Addressing this should be a priority if we are to reduce climate apathy. This means talking about the problem in terms of the impact it is likely to have on the places and things the swings know and value.

This isn't to say that the impact of climate change on people in other countries, and on nature, should be forgotten. Compared with people in richer countries, those in the poorest countries are likely to suffer more from global warming. The threats they will face will usually be more severe, and they are already mostly less able to cope with emergencies. It would be perverse not to mention them at all, both because the impact on them can't be ignored and because forgetting about the biggest victims of climate change – who also have the least responsibility for causing the problem – makes it harder to show why dealing with it is a moral imperative. What is more, focusing only on the impacts on richer countries risks legitimizing the argument that it is better to adapt to the consequences of global warming, rather than cutting emissions to limit it. Likewise, the impacts of climate change on nature are important both in their own right and because of the consequence for humans of the loss of other species.

Even so, describing climate change in terms of its impact in distant places doesn't seem to be effective with people who are starting from a position of apathy. The swings currently don't pay much attention to the threat. To start a conversation with the swings we need to talk about what climate change will mean for their countries. Chapter 10 discusses how we can build on this initial interest, to discuss the wider implications of warming and the moral imperative of cutting emissions.

As well as the focus on the distant impacts of climate change, another factor that makes it difficult for many people to understand why it matters to them is the impenetrability of the language used to describe the likely impacts of warming. Some of the terms we use to talk about how much the planet might heat up and what will happen when it does are difficult for most non-experts to understand. Even if someone does understand them intellectually, they lack an emotional punch. Our words will be more effective if we steer clear of dry concepts such as average global temperature increase ('four degrees') and annual sea-level rise ('one centimeter a year'). Since these are widely used ideas that can't just be dropped without good alternatives, and we have to come up with more engaging terms to replace them.

Before going on, there is a criticism of this approach that needs to be addressed. This chapter argues that one factor behind apathy is that many people don't realize what climate change really means for them, and that if they understood it better, they would become more concerned and willing to act. This sounds like the 'deficit model', which assumes people are driven by information but currently have limited knowledge. If they find out more, it says, their attitudes and behavior will change. But the deficit model is now generally seen as simplistic. As we have seen, most people's opinions and actions are based on a complex mass of motivations, some of which are rational and information-based, and some of which aren't. In addition, there's some evidence that fears of mortality, which knowledge of climate change can evoke, might even lead people to become higher-spending consumers,[4] suggesting that warning people about how global warming would threaten them could in fact be counter-productive. But while criticisms of the deficit model seem fair, they don't invalidate this approach because it is not intended to be taken in isolation. The argument set out in this chapter – about showing how climate change matters to people in richer countries – needs to be combined with other approaches, outlined in the next two chapters. If this chapter seems to focus too much on filling a gap in knowledge, that is because it is just one part of the overall approach.

So, with these challenges in mind, what kind of description of climate change could encourage more people to pay attention to it? The following description of the problem is focused on overcoming the specific problems that this chapter is seeking to address, that is, the widespread subconscious feeling that, since global warming is apparently neither visible nor local, it needn't be a high priority. Its intended audience is the swings, particularly the two relatively more engaged groups, the *Concerned* and the *Cautious*. To be able to use specific examples, the description is written for a UK audience, but the evidence could easily be adapted for people in other countries.

Over the course of writing this book, I have tested drafts of this description of climate change, and the equivalents in the next chapters, with people from both swing groups. Their responses have shaped the statements in ways that, they say, would make them more likely to support measures to address climate change. That isn't to say these paragraphs will be perfect for all swings – some people will be more persuaded with a different emphasis or different words. But my intention is to create descriptions of climate change that people who are worried about the problem can draw on when talking to those who are apathetic.

The first of the three statements is:

> 'People across Britain will suffer from dangerous climate change if greenhouse-gas emissions don't fall. In the last few years we have seen floods and storms that are early signs of climate change – but those were small compared with what could be ahead. Uncontrolled global warming would mean far more floods, from torrential rain and rising sea levels, which would devastate homes, businesses, railways and roads across Britain, including in places that never used to be at risk. In the summers we would face unbearable heat waves that would kill the old, the young and the ill. Temperatures would rise well into the 40s. And the effects of climate change in other countries would force hundreds of millions of people from their homes, with some trying to flee to Britain to escape famine and floods. Although we would be better able to cope than poorer places, not even richer countries like ours could shut ourselves off if the world's climate becomes this violent.'

Our fate, our hands

Since so many people unconsciously dismiss climate change, this description of the problem is intended to shift perceptions of it, from being something many people feel they should care about, but don't, to being something that more people recognize matters to them. The aim is to move global warming up the priority list.

This is why the description emphasizes what unchecked climate change would mean for people like the swings. In the UK, increased flooding is the threat that most people associate with climate change and this is the natural focus of such a description. As we saw in Chapter 6, heat waves aren't yet widely seen as a threat in the UK – the response to the temperature passing 25°C/77°F is usually excitement rather than alarm – but over time risks like these could become more commonly recognized. In places that are already hotter and drier, such as Australia and much of the US, threats like heat waves, droughts and wildfires are already well known; the 2016 wildfires in Alberta, Canada, reflected the fact that massive fires aren't restricted to countries closer to the tropics.

People are generally more likely to be concerned about climate change when they have personal experience of extreme weather that could be attributed to it. Such extremes have been described as 'benign catastrophes', meaning that, while they may be awful for the victims, at least they supposedly draw sufficient attention to climate change to spur action that will avert worse future disasters.[5] But there is not much evidence of this happening yet. It seems that weather-related catastrophes mostly affect the views only of the people who directly experience them, leaving wider public opinion apparently unaffected after the immediate crisis is over.[6]

Nevertheless, the understanding we all have of disasters like floods or fires, whether from direct experience or from news reports, means that they are the most convincing starting point for talking about climate change. Rather than referring to global degrees of warming or centimeters of sea-level rise, we should, wherever possible, talk about what those global changes would mean for our audience's local area – floods, fires, heat waves and droughts. In politics and marketing, the evidence that backs up an argument is sometimes known as 'proof points'. The consequences of global warming are the proof points to back up the argument for why we need to cut emissions. Indeed, while the description of climate change in this chapter is written at a national level, it may be more useful still if the proof points are local: telling people how their city, or even their street, might be affected by extreme warming would make it even easier for them to picture its effects.

This kind of approach has in fact been used by some climate campaigners. One campaign, 'For the love of', draws on evidence about how to describe climate change in ways that will persuade more people to care about the problem,[7] and focuses on how global warming will affect the things that, it believes, most people love. It uses examples – like 'my son', 'summer' and 'London', that matter to people who don't already have strong environmental values (as well as to those that do). While it also uses some particularly specific examples, like 'popcorn' and 'tea', which might confuse people in the swing groups who haven't already thought about how climate change could affect certain products, the campaign demonstrates that it is possible to talk about the impacts of dangerous warming in ways that show its local and immediate consequences.

The indirect effects of global warming, where changes happen in other countries and affect people in richer places, can also engage the swings, even if these effects aren't so easy to demonstrate. Compared with things like floods and fires, people who aren't already interested in climate change tend to find the indirect threats harder to believe. To people who are hearing about the connection for the first time, the link can sound tenuous and not credible. Debates about the refugee crisis caused by the ongoing Syrian civil war exemplify this. It is the kind of disaster that could demonstrate to many people how climate change might indirectly affect richer countries, as it seems likely that drought in the region, along with poor growing conditions elsewhere, was a factor behind the conflict, whose effects were eventually felt well beyond Syria's borders.[8] But media coverage of this research has often been met with incredulity, as if it is absurd that a complex, long-term and probabilistic process such as climate change, could possibly be a factor behind a sudden and devastating crisis like the breakdown of a major country, particularly when there were also other obvious causes that could be more easily blamed for the disaster.[9]

To overcome this, we should remember the conjunction fallacy, which we saw in Chapter 6, where people find it easier to accept the likelihood of a well-described event than a poorly described one, even though the well-described event is a subset of the poorly described one. We are telling a story. When talking about how people in richer countries would be affected by climate change in other places, we are likely to be more persuasive if we give examples of specific possible events with a clear chain of causality. In the Syrian example, just mentioning the first line ('climate change') and the last line ('the Syrian civil war') isn't enough. We have to build a satisfying connection that shows how the pieces fit together, taking our audience through each intermediate step. This is not only more accurate, it is more interesting and plausible for people who don't already know the details. Without it, they will be likely to dismiss the risk of events that are, in fact, quite plausible.

While the story of the indirect effects of climate change is more complicated and harder to convey, these effects shouldn't be overlooked when we are talking about how warming would affect richer countries. As well as further refugee crises, a similar argument could be made for the impact dangerous climate change would be likely to have on the prices and availability of food imported into richer countries. Not only do examples like these provide more evidence for why a changing climate matters to people in richer countries, but they also offer a route into talking about how warming would harm people in poorer countries, which is an important element of the moral argument for cutting emissions.

Some readers who are immersed in debates about climate change might be twitching by now, as so far there has been little mention of uncertainty – an aspect of the debate that is rarely left so late. Most discussions of climate change are punctuated with phrases like 'very likely' and 'medium confidence'. When strict accuracy is required, as in scientific journals, these terms are essential.

But when it comes to everyday conversation, they can be counterproductive and it is often clearer to discard them. Scientists have precise meanings for words like confidence, uncertainty and error, but most members of the public understand those words differently. When many people hear someone describing climate change in terms of confidence and error, they can easily think they are being told that the findings might be wrong. These phrases are used with the best of intentions – to make sure the descriptions of global warming are unimpeachable in their accuracy. But they can fail to convey the fact that the overwhelming majority of climate scientists have reached similar conclusions about the processes behind climate change and the consequences of continued emissions. Instead, they often give the impression that there is far more doubt than is warranted.

If we are determined to be absolutely accurate, at all costs, we would find it difficult to drop such nuances. After all, they are technically correct. It is never strictly accurate to claim certainty in science, since, by definition, science is never settled. So it is not quite true that human activities are *certainly* causing climate change or that increases in the planet's temperature will *definitely* lead to more energetic weather systems, even if the evidence overwhelmingly points that way. Indeed, not only is it inaccurate to describe knowledge about aspects of climate change as certain, it could, in some circumstances, be counterproductive. If a claim, made by a majority of relevant scientists with an assertion of total certainty, was eventually found to be wrong, it would only be fair to cast doubt on other things the scientists had said with similar apparent certainty.

But most of the time precision gets in the way of clarity and it is better to reach a sensible compromise between pedantry and everyday language. Where the overwhelming majority of relevant experts believe something is overwhelmingly likely, it seems unnecessarily obdurate to insist on referring to uncertainty, at least outside scientific forums. There is a helpful parallel in descriptions of smoking and health. Most people would agree that it would be ridiculous to object that the statement 'smoking causes cancer' is inaccurate, on the grounds that a few lucky people may smoke 20 cigarettes a day for 80 years without developing cancer. The reason those outliers don't invalidate the statement is that, in everyday language, the statement is understood as referring to probabilities. The word 'causes' is talking about the effect of smoking, on average, over a large number of people. No reasonable person would understand the statement to be claiming that every single smoker will, inevitably, get cancer. In the same way, it should be accepted as uncontroversial to say 'climate change causes floods'. It doesn't always need to be made explicit that, just as with the lucky smoker, not every part of the planet will be flooded more often as the world heats.

On the other hand, when a finding is tentative and disputed among fair-minded experts, it is right that such doubts should be made clear to the public. The loss of trust if apparently confident claims are proved wrong is too important. In fact, uncertainties are often neglected more than they should be in media coverage of climate research. Where a particular projection is made but is considered not to be particularly likely – for example, that there is a 33-per-cent chance that Arctic sea ice will disappear by a specific date – media coverage often focuses on the warning and drops the probability. This leaves researchers exposed to criticism if the event doesn't come to pass, which may be unjustified since they never said it would definitely happen. This was the fate of the UK's top weather forecasting body, the Met Office, which came under attack in 2009 when it projected a 50-per-cent chance that summer temperatures would be above average. The media coverage, albeit encouraged by the Met Office's own press release,[10] referred only to the prediction of a 'barbecue summer', neglecting to mention that the projection had found there was a 35-per-cent change that the summer would be too wet for grilling meat.[11] When the barbecue summer never arrived, the Met Office was roundly ridiculed. It was even referred to in Parliament, by an MP (who was no climate denier) who described long-range forecasts as 'discredited' because of the Met Office's supposed prediction.[12]

But despite the risk of relatively low-probability projections being overhyped, discussions of climate change would mostly benefit from dropping the cautious language of uncertainties and probabilities where there is little serious scientific doubt about the basic principle. The marginal benefit gained from being technically accurate doesn't outweigh the misunderstanding and overestimation of doubt that these terms engender. And where uncertainty is too important to be overlooked, there are often better ways of talking about it. One option is to refer, not to uncertainty, but to risk – a more emotionally engaging term that triggers most people's desire to avoid losses. Scientists who don't feel comfortable dispensing with references to uncertainty and error margins, even when talking to the public, could – if they prefer – discuss their findings in those terms and then add a personal reflection of how the results make them feel. Many people who would be unmoved by talk of probabilities, may respond to hearing a scientist say that their research into Arctic sea ice had left them scared for the future.

Other simple changes of emphasis can be useful. For example, instead of talking about a range of possible consequences of global warming, it may be clearer to focus on a specific change and acknowledge that scientists don't know exactly when it will happen. So, instead of saying that, in 2060, a particular place will face between 80 and 120 days a year that are hotter than 35°C/95°F, we could say that sometime between 2050 and 2070 it will have a year with 100 days hotter than 35°C/95°F for the first time. The latter phrasing is more natural and so easier to relate to everyday life. Climate Outreach, which studies public responses to climate-change communications, identified several similar ways of talking about uncertainty in a 2015 handbook.[13]

While the description in this chapter of why climate change matters is focused on the effect extreme warming would have on people in richer countries, it also alludes to the world's ability to deal with the problem. With its emphasis on the consequences of the world not cutting emissions, it suggests that there is still a chance of avoiding this disaster. There are two sides to this. The optimistic side – what the world can achieve if it focuses on averting dangerous warming – is discussed in Chapter 10. The pessimistic side – whether it is worth the effort – is tackled in the remainder of this chapter.

Recently, some of those who previously claimed there was no evidence of a link between human activities and climate change have quietly shifted their emphasis. Instead of denying that climate change is caused by humans, or even that it will be a problem, some have begun trying out the argument that, yes, climate change is happening, and, yes, it may be a problem, but in fact efforts to cut emissions have failed and we in richer countries should focus above all on protecting ourselves from it. According to this logic, reducing our country's emissions is pointless as it is too late to prevent global warming; instead, all our energy should go into strengthening our defenses against floods, heat waves, droughts and wildfires. Ultimately, this suggests all our country should do is wait and prepare as best it can for what will hit it in the coming decades. But this doesn't hold up to scrutiny.

There are two way of looking at the question of whether it is truly too late to prevent climate change. On one side, the level of emissions the world produces over the next couple of decades will determine the extent of warming the planet faces in the lifetime of many people alive now, and certainly that of their children. So there clearly is still time to influence climate change. If the world could somehow stop producing all emissions overnight, the amount of warming it will face would be vastly reduced. But, on the other side, people worried about climate change often emphasize the urgency of decisions being made now, sometimes giving a figure of months left to save the planet.[14] When this is then combined with regular denunciation of the world for being too slow to deal with the problem, an observer who heard these arguments might conclude that the chance to save the planet had been missed and it is now doomed.

A solution to bridging these two sides is to stop presenting climate change as an all-or-nothing problem. Since the world has already warmed by nearly 1°C/2°F as a result of human activities, it is clearly wrong to suggest that climate change can be entirely prevented. And some further climate change cannot be fully prevented either. If emissions were halted overnight, the warming may soon stop[15] – but that isn't going to happen. There is already a world of power plants, heavy industry and transport systems that rely on fossil fuels, and it won't be switched off suddenly. If all of those systems were used until 2060 and no more polluting machinery was built – already a near-impossible scenario – the world would warm by about half as much again as it already has.[16] So it is not possible to prevent climate change entirely. But what the world can do is determine how much more warming it will face: either an amount that is manageable without unprecedented disruption (if decent support is in place for the world's most vulnerable people, including resettlement for the people whose homes, land or entire country is lost to the sea); or an amount that will be catastrophic for billions of people and for economies worldwide. The world can still stop dangerous global warming.

There is no escaping the fact that this is an unhappy outlook. Whatever the world does now, it will face a less hospitable climate in the next few decades. This is a problem for overcoming apathy, as there are limits to how far people can be motivated by a call to achieve the lesser of two evils. So, while showing the impact that climate change will have on people in richer countries is valuable for widening interest in it in the first place, it is unlikely to be sufficient to persuade many people to change how they live. The final chapter discusses ways we can talk about climate change to build enthusiasm about what is possible.

The other side of the argument for why people in richer countries supposedly needn't worry about cutting emissions is the suggestion that they can adapt to any changes that come their way. So even if it were widely recognized that it is still within the world's power to determine the extent of warming it faces, there is still a risk that any country with high emissions might think it better to focus on protecting itself against the effects of a worsening climate, rather than reducing its contribution to the problem. In the Paris Agreement, governments committed to limiting climate change, yet the deal doesn't include any means of international enforcement. The risk of a country free-riding on the efforts of others has usually been seen as unlikely, because it is assumed that other countries would shame them internationally, or would follow suit, precipitating the collapse of the deal entirely – which wouldn't be in the interests of the free-riding country. But that may not bother a powerful country led by a government that was unconcerned about climate change or even actively hostile to efforts to reduce fossil-fuel use. Yet although international agreements can't compel governments to cut emissions, it is still a mistake for any country to believe that adapting to a harsher climate is a viable alternative to seeking to prevent that deterioration.

People in poorer countries would be the worst affected by severe climate change, but even those in richer ones would face changes they couldn't easily adapt to. No country with a large coastline and tidal rivers could build high-enough walls to prevent rising sea levels from regularly flooding homes and businesses. The same goes for attempts to deal with storms that are more devastating than anything seen before, rivers that flood more often, heat waves that kill thousands and wildfires that destroy great areas of land, or the fallout from disasters in other countries. Some countries might be able to protect some of their richest cities and a few lucky towns, but, as the failure to save New Orleans from Hurricane Katrina in 2005, and the impact of Hurricane Sandy on New York in 2012 proved, even the richest country in the world has so far been unable to build defenses that are sufficient in even the current climate to stop major cities from being overwhelmed. Richer countries can't relax about global warming on the basis that they will just be able to deal with as the climate worsens. Averting extreme climate change offers a much better outlook than they would have if they tried to live with it.

This chapter's description of how we can persuade swings that climate change matters has focused on showing that it affects people and places that are, in Wensleydale's words, 'really *real*'. Discussions of the impacts of climate change haven't yet shown most people in richer countries, particularly the people who are apathetic, how it will affect them. If we can earn their attention, we can go on to talk about the other consequences of climate change, but such consequences are unlikely to interest many in joining the conversation in the first place. As well as showing why global warming matters to people in richer countries, we also need to convince them that the world still has the power to decide the kind of climate it will face in the future. The decisions it takes now will shape whether or not the world will have to live with catastrophic global warming, or whether climate change is limited to more manageable levels. Together, these should attract more people to talk about climate change. The next chapters look at what else should be part of those conversations.

9
Tear down this wall

Discussions about climate change should give more prominence to people who are not seen as leftwing, and should encourage high-profile debates about possible solutions. This is essential if more people from the Center and Right are to be persuaded to take the problem seriously.

'I would identify more with the "we" if there was a picture of people who you didn't identify as environmentalists, who just looked quite ordinary.'
– Focus group participant, Climate Outreach, UK[1]

The aim of the last chapter was to show how we can begin to interest the many people who accept that climate change is a problem but still don't think much about it. Yet even if far more people realized how extreme warming would affect them, their family and the places they care about, it is unlikely that this alone would be enough to persuade them to support the actions that might be needed to prevent it. Increased understanding of the consequences of climate change is necessary for building this support, but it isn't sufficient. Other factors behind climate apathy also need to be overcome.

This chapter focuses on the way that climate change is seen as an interest of the political Left, particularly environmentalists on the open Left, and has become a battleground between Left and Right. This has contributed to climate apathy in various ways. First, descriptions of why global warming matters often focus on the environment and international justice. This appeals to people with a particular set of values, but these people are generally already worried about climate change. They are mostly in the base segment and, while their activism will still be needed, they don't number enough to form a majority. For people in the swing segments, evidence about the impacts of climate change on animals and distant places is unlikely to be enough.

Second, many people, including most of the swings, don't identify with environmentalism and the open Left. They might not be trenchantly opposed – the swings tend to be close to the middle of the political spectrum – but they nevertheless mostly see themselves as having different political values. As a result, they are more likely to regard climate change as an issue that is of interest to different kinds of people, and not as something that people like them normally care about.

And third, the association between climate change and the open Left encourages some to question the motivations of people advocating emission cuts. There is an unresolved debate about whether growth-focused capitalism can be reconciled with effective action to limit warming. But regardless of whether they are truly inimical, some of those who say climate change is a reason to make revolutionary economic changes are sometimes accused of exaggerating the threat to argue for redistribution they already wanted, regardless of global warming.

Together, these factors make some swings suspicious of what they hear about climate change. When evidence-based arguments are made about how warming could affect rich countries, the polarization of the topic means that many people are suspicious of what they hear. In the terms of Zaller's Receive-Accept-Sample model, discussed in Chapter 5, swings may receive the information but often they don't accept it. Overcoming this suspicion is the focus of this chapter.

Everyone's problem

The recent history of European politics offers a cautionary tale for anyone trying to shift how their cause is viewed.

In the face of the decline of the industrial working class, center-left parties in Europe found themselves in trouble. From the 1970s, as economies deindustrialized and trade-union membership fell, it became increasing clear that there were too few traditional working-class voters for parties of the Left to win power principally on the basis of their support – even with the help of leftwing members of the middle class. In some countries, the main leftwing parties responded to the change by repositioning themselves as more centrist, with a platform influenced by Bill Clinton's presidency, designed to win over centrist middle-class voters while maintaining enough appeal to working-class voters and leftwing members of the middle class to avoid losing their support. In electoral terms it worked brilliantly, with parties like Germany's Social Democratic Party and the UK's Labour Party winning power from 1998 to 2005 and 1997 to 2010 respectively.

But as time went by, those center-left parties that had gone after centrist voters began to struggle to answer the question, Why do you exist? This became even more acute after the financial crisis made it harder for governments to expand state spending in the way center-left governments had done in the boom years, when they had been able to offer improved public services to their traditional supporters while reassuring more affluent voters that they could do so without being seen to raise taxes. With public finances tighter, it was no longer so easy for these parties to sustain the coalition of their traditional supporters and more conservative voters. Many of their old supporters had become disillusioned with center-left parties' claims to be their voice, while conservative middle-class voters became more doubtful about the parties' ability to govern competently. The old parties of the Center-Left struggled to speak for any group, with the result being the dominance of European politics for the years after the financial crisis by rightwing parties.[3]

The fate of the European Center-Left is useful to keep in mind when thinking about how we talk about climate change. At the moment, advocates of strong action to limit warming have the secure backing of the base segment – around one person in five. It is a minority and it doesn't seem likely to be enough for the toughest measures that will be needed to cut emissions. But it is still better than nothing, and the cause of averting disaster would be set back enormously if efforts to win over more people ended up not only failing to persuade the swings but also alienating those who already care. With this in mind, the challenge must be not only to persuade the swings that climate change matters to them, but to do that without undermining the concern and energy of those who are already committed to tackling the problem.

The previous chapter was focused on what we could say about climate change to interest the swings; this chapter focuses on how we can say it. It discusses the context of conversations about global warming, with the aim of increasing the chances that someone who hears about the issue will accept what they hear. The argument is again based around a description of climate change, which can be summarized as:

> 'Uncontrolled climate change would be a problem for everyone – not just environmentalists. It is often portrayed as something people are only interested in if they are leftwing or green. But the world won't stop dangerous climate change if the solutions only come from people with one set of perspectives. Fortunately, people from across society, with a range of political views, agree that climate change is a serious threat and that we need to work together to stop it. This includes leaders of some of the world's biggest businesses, like Coca-Cola, Unilever and Nestlé, the British army and our NATO allies, and politicians from across the UK's main parties. These people often disagree with each other about how to deal with the problem, but those disagreements are a good way of working out the best answers.'

More than the other descriptions of climate change in these chapters, the argument outlined in this paragraph needs to be shown to be true. Some swings I spoke to responded to drafts of it with polite disbelief. They thought the principle of climate change being a cross-partisan issue was a good one but saw little evidence that it is true. For the reasons discussed in Chapter 7, this isn't surprising. The challenge is to *show* that it is true, not just to assert that it is.

Equally, the argument may be uncomfortable for some people. Some may hold the view that climate change is an opportunity to boost long-standing arguments for socially progressive economic revolution. Others might believe that stopping catastrophic climate change and continued economic growth are irreconcilable, so economic revolution is essential if the world is to escape disaster. Supporters of either view might object to the argument that the widespread association between climate change and the open Left should be broken. But if we are serious about wanting to avoid disastrous warming, such a break may be essential.

Efforts to win widespread support for radically cutting emissions are hindered by the argument that the issue should be subsumed into the wider struggle against global injustices. This is not to say other injustices are less deserving of attention. But it seems unlikely that adding climate change to the struggle against those injustices would add a great deal to their cause and it would be likely to diminish the attention given to global warming. Most people are not passionate advocates for global justice. If they hear about climate change as one more argument for a fairer world, we shouldn't be surprised if they fail to appreciate the extreme urgency of cutting emissions. Most people would see climate change as, at best, another important but long-standing problem – one that should be addressed some time but never needs to be the highest priority. Making climate change an argument for tackling other injustices risks diminishing its importance in many people's eyes. Ultimately, this increases the danger that the world's climate will become more violent, which particularly hurts those who already suffer the most from injustices.

What about the argument that radical economic change is essential, not only because it is needed to tackle other injustices, but because without it the world will never avoid catastrophic warming? While world economies are now cleaning up sufficiently quickly to balance out economic growth – so global emissions largely stopped increasing from 2014 – it is not yet clear whether this will continue and whether it can be taken further so annual emissions start falling by enough to avoid dangerous warming (don't forget how steeply emissions need to fall to prevent dangerous warming, as shown by the chart in Chapter 3). In fact, the world might have already made the easier emissions cuts, and so, as economic growth continues, emissions could start increasing again.

But there are reasons to believe the world could, in fact, significantly cut emissions without ending economic growth. Electricity production can be made radically cleaner, with every coal power station shut down and an enormous expansion of renewable power that is combined with storage for times when the sun isn't shining or the wind isn't blowing. This switch is already starting to happen, with the cost of renewable power falling rapidly – the price of solar panels fell by 80 per cent between 2009 and 2015[4] – it is sometimes now the cheapest option.[5] This has been fostered by government incentives and regulations, which allowed new industries to gain experience and cut costs without being exposed to established competitors, but within decades global electricity production could become close to carbon-free purely through market forces. Investment in solar-power production may offer a reliable rate of return for investment funds looking for a long-term profit, suggesting clean energy could thrive in a growth-based model.[6] In other areas, like transport and heating, the same process could apply, with governments initially using incentives and pollution rules to develop clean alternatives that can be a source of profit and growth. Governments could also create markets to absorb greenhouse gases, with the technologies discussed in Chapter 3. None of this requires economic growth to stop – the move to low-carbon technologies described here could be spread by companies that are seeking to sell their products for a profit – and offers the potential for a much faster rate of decarbonization than the world has seen so far, suggesting that capitalism and the climate are not inevitably irreconcilable.

This is a compromise. Neoliberals may not like the use of government interventions in markets to put a price on pollution and protect clean alternatives so they can develop and eventually compete. Equally, the task of cutting emissions would be greater if economies continued to grow than it would be if economic activity declined. And this is to say nothing of the other dangerous effects of continued economic growth, for example the loss of biodiversity, the build-up of waste in the environment and the erosion of soils around the world. There are similar debates about whether those looming disasters can be kept at bay within growth-focused economies, and one of them may indeed provide an unchallengeable reason for halting growth.[7] But, when it comes to climate change, even if we believed that the best that can realistically be achieved, after growth, is something like the middle scenario described in Chapter 4 (and it may still be possible for the world to do better), that would still be much safer than the warming the world would face if it gave up entirely on cutting emissions.

Why should we accept that compromise, if it means the world has to do more to reduce emissions? After all, capitalism is surely behind the climate crisis – without fossil-fuel-based economic growth fostered by the search for profit, the world wouldn't have an emissions problem. The answer is that the task of persuading the public to switch to a zero-growth economy would add an immense challenge. It would require radical changes to society and would have no chance of happening without a clear majority behind it. Winning this support depends on persuading many people who are not part of the open Left that such a revolution is necessary. Even with climate change providing another reason why abandoning growth may be beneficial, it is difficult to see such a majority being formed in the near future. Success wouldn't mean everything is resolved, either. The world would still face the challenge of hugely reducing its emissions – halting growth would lessen the task but, even with economic activity frozen at current levels, cutting greenhouse gases would still be an enormous task. And there are costs to trying and failing. If we put all our efforts into arguing that climate change requires an end to economic growth, we might marginalize the cause still further and lose the time that could otherwise be spent finding ways to cut emissions now.

You might still think that an end to economic growth is the only way the world has any chance of avoiding catastrophic warming. If so, the challenge of overcoming public opposition would remain a lesser problem. But whether or not you think radical economic change is essential for avoiding dangerous global warming, winning support for such a revolution – as for significant cuts in emissions – surely depends on engaging people who don't associate themselves with the open Left. Linking efforts to cut emissions with campaigns that are identified with that group, such as the environmental movement and efforts to revolutionize global economies, is unlikely to help and, if possible, it seems worth breaking the association.

A good argument

If our aim is to change the view that global warming interests only environmentalists on the open Left, an obvious solution would be to make sure people with different political views are also seen to talk about it. If, instead of concern about climate change being apparently limited to environmental activists and others with similar views, the issue was seen to worry apolitical people and those from the Right, perhaps it would lose its association with the open Left.

But, as we saw in Chapter 7, that is already happening. There is no shortage of people from beyond the open Left who publicly warn about the risks of climate change and call for action to address it. Yet most members of the public who aren't already interested in the problem seem not to be moved by these warnings. Compared with the fabricated arguments about whether or not climate change is caused by humans and whether it is a threat, and the occasional protests by environmental groups that are imaginative enough to get media coverage, the pronouncements by businessmen (they are usually men) make for boring news stories that are, at best, relegated to the financial pages.

If the ways in which the political Center and Right, and the apolitical, talk about climate change don't get much attention, what can they do instead? One option might be for them to turn up the volume – instead of having a few isolated business leaders warning about climate change from time to time, more of them could do it, more often and more insistently. A coalition of leaders of the world's top businesses making regular public calls for significant measures to avert climate disaster might command attention. But it still wouldn't provide much conflict and may not be able to compete with the aspects of the issue that the media prefer – the arguments about science and the occasional high-profile protest.

Instead, if people who are worried about climate change, but aren't part of the open Left, want more people to pay attention to them, they need to become more interesting. The coverage of a gaggle of top CEOs chaining themselves to a coal-fired power station would be entertaining, if rather unlikely to happen. But what could make a great deal of difference to how climate change is viewed is for them to change the subject.

Climate-change websites, journals and conferences are filled with arguments about how the world could cut emissions, but most people don't see them there. Yet these arguments offer a way for people from across the political spectrum to give the media the conflict they need for high-profile coverage. Unlike the climate-change disputes that the media tend to cover, these disagreements are between people who have no doubts that global warming requires urgent attention. They focus on what the world should do about the problem, not on whether there is a problem that needs to be addressed.

There are plenty of controversial disagreements that could have a higher profile. Under the Paris Agreement, governments from around the world have already committed to certain emissions cuts. Over the coming years, each country is due to offer stronger cuts, but there has been little public debate about how each country should determine its contribution. The default option is to base it on their current emissions, but that seems unfair on poorer countries. An alternative would be to follow a principle of allowing each country's limit to be based on its population size. This seems fairer but is hard for high-emitting countries to swallow. This debate – which was temporarily settled by the Paris Agreement but will return – is rich in controversy, and thoughtful, well-informed people could disagree with one another, without ever disputing that emissions cuts are needed.

Within a country that is aiming to cut its emissions, there is scope for many similar debates. For example, if a community doesn't want a proposed local power plant, such as a wind farm or solar array, should they be entitled to veto the development? The democratic answer may seem to be that they should, but the consequence of a veto might be that a more expensive development is built instead, such as an offshore wind farm. If such an approach was widespread, energy bills would be more expensive for everyone, for the sake of a few communities. Here, too, there is room for higher-profile arguments among people who have no doubts about the threat.

Similarly, nuclear power divides those who are worried about the climate. Some argue that it is the only technology that doesn't produce a large volume of greenhouse gases and can be counted on to produce electricity on a large-enough scale and with enough consistency to replace power plants that burn coal and gas. But some environmentalists are appalled by nuclear power, seeing it as no improvement on coal. This is another conflict-ridden debate that can demonstrate to wide audiences that well-informed people can disagree about energy policy, without any of them disputing that climate change requires attention.

And when it comes to preparing for the effects of climate change, there are many other controversial areas that are rarely given mainstream attention. One question relates to the protection of land. If an area that is currently being used for housing, farming or public infrastructure starts to flood much more often than before, as a result of climate change, whose responsibility is it to improve flood defenses and to repair the damage? It may be that insurance companies can cope in the short term, but eventually that might be unsustainable and it may be necessary to abandon some areas. If so, there will be difficult decisions to be made about the extent of the government's responsibility to protect all parts of the country, and how far future flood risk should be taken into account when planning permission is being considered.

What these potential arguments – and the many others like them – have in common is that they provide conflict about climate change that could interest the media, without depending on disagreements about whether global warming is real, or on only using voices from the open Left as advocates of action to cut emissions. Even if some of those debates were to become polarized on the Left-Right axis, the existence of the debate would still show that people from across the political spectrum consider it a serious threat.

These arguments aren't just helpful for showing the diversity of people who are worried about climate change; they can also improve the quality of proposals for dealing with the problem. Debates about nuclear power are a good example. A UK poll found that those who think we don't need to worry about the climate for now are 50 per cent more likely to support nuclear power than those who think we should act now to prevent climate change.[8] Since nuclear power is one of the limited tools that could create a constant supply of low-carbon electricity on a large scale, this is surprising. While some environmentalist opponents of nuclear power say that its cost is their main objection,[9] this isn't every opponent's principal motivation. Many are at least as concerned about risks of radioactive pollution and the technology being used to develop nuclear weapons. But because of many environmentalists' passionate opposition to nuclear power, debates about it are typically overheated. Having a more open debate about nuclear power, among people who share the view that emissions need to be cut radically, would help to explore and test objections to it and balance these objections against the alternatives.

It is easy to underestimate how large a step it would be for many environmental activists to allow the debate to be widened in this way. As reflected by the accusation, made by a prominent environmental writer, that support for the expansion of nuclear power is a 'new form of denial',[10] debates about responses to climate change can be vituperative. It isn't self-evident to many environmentalists that a shared belief in the urgent need to cut emissions is enough reason to co-operate with people whom they otherwise strongly disagree with. But, if the swings are to be persuaded that climate change isn't just a leftwing issue, it seems essential.

A movement called Ecomodernism provides an interesting example of the potential and the risks of this kind of approach. *The Ecomodernist Manifesto*, launched in 2015, tried to reconcile economic growth and protection of the planet. Its aim was to show that this reconciliation is possible if human activities are steered to interfere less with the natural world, allowing humans 'to mitigate climate change, to spare nature, and to alleviate global poverty'[11] – that is, to keep expanding economies while reducing emissions and other impacts on the environment. Among the authors of the manifesto were several high-profile environmentalists, yet its approach was sharply different from that of many other environmentalists. The launch of the manifesto prompted criticisms from many who doubted the benefits of the Ecomodernist vision of a world, where the environment is best protected by humans separating themselves from it. The critics argued that seeking to end small-scale farming would push billions of people into poverty in urban slums.[12] But while this argument may have shown environmentalists to be divided, it also demonstrated to a wider audience that there are people who think climate change is an urgent problem and who don't hold other views associated with the open Left.

Yet the manifesto's launch also showed how this approach can fail. One of its lead authors, Mark Lynas, described it as 'a screw-up of impressive proportions'.[13] The Ecomodernists' mistake was to pitch too large a tent and to allow themselves to be associated with politicians who were widely considered to be climate deniers. Predictably, when the media covered the manifesto's launch, they did so by fitting it into the traditional frame of a dispute about whether climate change is a significant problem. So while Ecomodernism's launch demonstrated that it is possible to get high-profile media coverage that shows a diversity of views among people worried about the climate, it also reflected that journalists have become used to covering the issue in a particular way and it can be a challenge to persuade them that there is another, more interesting, story.

Different kinds of people

Changing the focus of controversies about global warming from whether it matters to what the world should do about it may be useful for showing that it is not just an issue of the open Left, but it isn't the only way to do that. Another approach is to change who is seen to be talking about the problem, so more swings hear about it from those they trust.

There are two kinds of people who would be particularly trusted to talk about climate change and currently aren't heard as often as they should be. The first is scientists. As we saw in Chapter 2, scientists are still the most widely trusted to talk about global warming, despite the attacks that have been launched on them in recent years, particularly in the US. If our aim is to convey information about climate change in ways that more people accept, drawing on this trust in climate science is essential.

But how far can this go? The trust most people have for climate scientists is based on the understanding that their work is, as far as possible, an impartial search for information. People expect them to talk about the evidence that it is happening and why it is happening, and how it will affect the world if it is not addressed. When it comes to talking about possible responses to the problem, the ground is more hazardous. Opponents of climate science often accuse scientists who do this of having become activists. Even so, when they talk about responses to climate change, scientists can draw on a reservoir of trust that they – nearly alone in the debate – have access to. A useful distinction may be for scientists to make clear when they are going beyond their research findings, and are speaking as an individual about what they think is needed. Opponents may still claim that this shows the scientists to be activists, but most reasonable people would recognize that the person who dedicates their life to uncovering evidence about climate change should also be heard when they talk about what they believe the world should do about it. Not all scientists want to speak publicly – after all, they chose to become researchers rather than campaigners. But those climate scientists who are willing to talk about their work should recognize that they are unusual in having this valuable resource of public trust.

The other group who can talk about climate change in ways that would be more likely to be accepted is ordinary people. This might seem to contradict the evidence on public opinion – while 69 per cent say they trust scientists to give them reliable information about climate change, only 30 per cent say the same about their friends and family.[14] But in reality people are most likely to be persuaded to change their views or behavior as a result of influence from people they know and trust. Perhaps the explanation for the low trust in what friends and family say about climate change is that it seems a complex and contested area and most people think others around them don't know that much about it. If an individual can persuade their friends and family that they do in fact know what they are talking about when it comes to global warming, they have a good chance of changing minds.

If ordinary people, scientists, and those from across the political spectrum all have the potential to help address the problem that the swings generally don't pay attention to climate change, what about environmentalists? Since part of the problem is that climate change is seen as an issue that interests the open Left, perhaps environmentalists should talk about it a bit less, to leave space for others.

But while doing so might shift the view of climate change away from being associated with the open Left, it would be counterproductive. Environmental activism does an enormous amount of good, and abandoning it would be a significant loss. Targeted, noisy and trouble-making activists have been behind many of the decisions that have helped the climate in recent years, like measures to limit oil and gas exploration in the Arctic, President Obama's rejection of the Keystone XL Pipeline and the closure of coalmines and power plants around the world. Companies and politicians targeted by activists often conclude it is better to concede than to fight on, even if the activists are just a small, if well-organized, group. While many of the changes needed over the coming years are beyond what activists can achieve without wider support, activism has driven the world's recent progress in slowing the growth in emissions and it is likely that it will continue to be essential, particularly if some governments start trying to undo recent progress, as with President Trump's revival of the Keystone XL project in January 2017.

A resolution to this tension – that environmental activism might motivate the base and achieve important victories, but put off the swings who will be needed for future battles – is to recognize that different messages, arguments and messengers appeal to different audiences. Some of the large environmental campaign groups have members who are mostly in the base segment. It is unreasonable to expect these groups to do and say things that would antagonize their members, and their role may always be to mobilize the most enthusiastic activists. But other organizations – including some businesses and more conservative environmental groups – and people can talk to the swing groups in language that appeals to them. The challenge for everyone worried about climate change is to make sure that the right voices are heard in the right places.

A further reason the open Left should carry on talking about climate change is that they might be able to produce fundamental shifts in public attitudes. A particular model of understanding decision-making – which holds that an individual's behavior is determined by their values – is influential among many people worried about the climate. The Common Cause Foundation, a prominent advocate of the model, categorizes values into two competing sets: intrinsic and extrinsic.[15] Everyone has values from both sets, but people whose values are predominantly intrinsic tend to be principally concerned with problems that go beyond themselves, such as their community, family and friends, and their self-development. On the other hand, people with predominantly extrinsic values are mostly interested in how they are seen by others, and are more likely to prioritize wealth and power.

While it is possible to argue for some actions to tackle global warming from an extrinsic approach – for example that an energy-efficient home would be cheaper to run, leaving more to spend on other high-status things – supporters of this model believe that vital aspects of the response to climate change can only be advocated on the basis of intrinsic values. For example, it may be hard to persuade many people to fly less often without drawing on values that go beyond the self. Significantly, they also suggest it is possible, through a deliberate effort, to shift the balance between intrinsic and extrinsic values in individuals and ultimately in society, so that more people could be persuaded to prioritize problems that go beyond themselves.

This last part is disputed. Research evidence backs up the claim that it is possible, at least in the short term, to shift an individual's values towards more socially beneficial ones.[16] But some people argue that it is too great a challenge for any campaign deliberately to produce a measurable shift in the balance of the public's intrinsic and extrinsic values. They suggest it is better to work with people's existing values, not only because it is difficult to shift values, but also because an argument that is based on alien values is unlikely to be welcome.[17] Considering how much information bombards an individual from advertising, news, and the conscious and unconscious signals from the people around them, it certainly seems like an enormous challenge for any organization to cut through the noise and measurably move the values of a significant number of people.

And yet, if the theory is correct and it really is possible to change society-wide values, the prize would be enormous. A campaign that shifted the public towards intrinsic values could transform attitudes to the climate, as well as to other issues. Giving up on that possibility might be not only a missed opportunity but also a risk. It is nearly impossible for an individual to avoid a stream of advertising that claims wealth and consumption bring happiness. Without an effort to shift public values towards the intrinsic, there might be no resistance to these extrinsic messages. So, while there are downsides to climate change being too closely associated with the environmental Left, there are also risks with that side of the debate withdrawing entirely.

This chapter set out to find solutions to the problem that many people struggle to accept what they hear about climate change because they see it as an issue that interests only people with a particular set of political values. Part of the answer may be to demonstrate that the debate is wider than it often seems. It is already the case that far more than just the open Left talk about climate change, yet these other voices are rarely heard by people who aren't already paying close attention. Scientists and ordinary members of the public also have the potential to show that concern about climate change goes beyond one corner of the political spectrum. Many of the debates about responses to the problem could provide the conflict the media need, while demonstrating this diversity among people who worry about the issue.

Running through the chapter has been concern that climate change could go the way of the European Center-Left, alienating its traditional supporters without motivating many other people. Although the association between climate change and the Left brings problems, divorce isn't the answer. Climate activists have won repeated victories and their work will be needed for years ahead. No-one else is likely to be so fast or effective at applying pressure where it is needed, even if there are times when other voices are likely to be essential.

Along with the previous one, this chapter has focused on the damage that extreme climate change would cause and how this can be conveyed to people who are currently apathetic about the problem. But while this may be necessary for getting more people interested in the issue, it won't be sufficient to motivate change. The final chapter looks at what else is needed: how can we build a positive case for tackling climate change, which motivates more people to want to deal with it, rather than just trying to frighten them into it?

10
Yes we can

To overcome apathy we also need to show that cutting emissions will bring benefits beyond avoiding dangerous climate change. Leaders need to prove that the burden is distributed fairly and that dealing with the problem is not just self-interest but a moral imperative.

'The whole fury and might of the enemy must very soon be turned on us... If we can stand up to him, all Europe may be freed and the life of the world may move forward into broad, sunlit uplands. But if we fail, then the whole world, including the United States, including all that we have known and cared for, will sink into the abyss of a new dark age.'
– Winston Churchill, June 1940[1]

Scaring people about climate change isn't enough. Even if far more people understood how uncontrolled climate change would affect them, and became more concerned about it, there would still be no guarantee that they would be prepared to make sacrifices to deal with it. In fact, relentless discussion of the dangers of global warming could encourage many people to try to stop thinking about it. Instead of focusing attention, fearmongering can be counterproductive.

This chapter aims to avert that risk. While talking about the threat that climate change poses can help more people see why it matters to them, conversations about the problem shouldn't stop there. We also need to talk about the benefits of cutting emissions. We can show that extreme climate change can be prevented and that doing so will allow the world to improve lives and communities in other ways. And, after establishing more interest in the problem and demonstrating that solutions are possible, we can move beyond a narrow focus on how global warming would affect the lives of people in richer countries and begin to foster solidarity with the billions of other people who will also be affected.

Climate change is a scary and unwelcome consequence of human activities. We would be misleading ourselves and others if we claimed that it is good news. Some gloom is justified and too much optimism could undermine our case that it deserves attention. But if there is no hint of sunlit uplands ahead, it is not surprising if many people prefer not to think about it. The challenge, and focus of this chapter, is finding the appropriate balance between optimism and fear.

Collective responsibility

A scene in the TV show *Mad Men* neatly reminds us that public behavior can change for the better. In the episode,[2] the characters enjoy a picnic in a beauty spot before returning to their car, leaving their trash on the ground. The audience is prompted to reflect on how attitudes have changed since 1962, when the episode was set – most people now wouldn't dream of casually leaving their garbage behind, particularly in such a beautiful place, yet the characters do so without a backward glance.

From the characters' perspective, such behavior might seem rational. Taking the trash away would be an effort and, if they weren't planning on coming back soon, they wouldn't be inconvenienced by the spot being garbage-strewn. There was no serious risk of punishment – even if dropping trash was considered unacceptable at the time, they could see that no-one was around to call them out for it. It seems to be an example of a 'Tragedy of the Commons', Garrett Hardin's 1968 theory[3] that, without proper oversight, individuals overuse communal resources (in this case, a pristine environment) to the detriment of everyone else.

This doesn't seem to be an unfair characterization of behavior in that era – a 1968 study found that 50 per cent of Americans admitted to dropping trash. But attitudes have changed and research in 2008 found just 15 per cent said the same.[4] So even though discarding your trash can seem rational, most people now say they wouldn't be so relaxed about it. An individual knows their discarded trash wouldn't make much difference to the overall tidiness of their country – and the chance they will be caught and punished is slim[5] – yet most people still go to the effort of disposing of it properly. It seems most people have, on the whole, decided that they prefer to live in an environment that is trash-free and that the way to do that is to behave as they would expect others to. Formal punishment for dropping trash is unlikely, so most people appear to be kept in check by their own moral codes or by fear of disapproval from others.

A similar attitude of collective responsibility is essential for climate change, not only at the level of individuals, but also at that of countries. One of the common arguments used to suggest an individual country shouldn't bother cutting its own emissions is that doing so would make little difference to the global total. Australia, Canada and the UK's emissions each represent only around one to two per cent of the world's total.[6] In any of those countries – as in nearly every other – no amount of emission cuts could make a large direct contribution to global efforts to avoid dangerous warming. It is only China and the US that could cut much larger amounts individually – around 30 per cent and 14 per cent respectively[7] – and even in either of those countries someone could point to the rest of the world and say action is of little use if others do nothing. Widespread freeloading would spell the end of emissions cuts.

Fortunately, pledges by nearly every country to limit their emissions,[8] made for the 2015 Paris Agreement, demonstrate that individual countries are willing to act, regardless of how large a proportion of world emissions they are responsible for. But the existence of this global commitment to cut emissions appears not to be widely recognized. Critics of efforts to cut emissions still argue that individual richer countries shouldn't go to the effort of cutting their emissions because it wouldn't make much difference globally, and, it is claimed, other countries aren't making a similar effort ('it's okay to leave our rubbish because it will soon blow away and other people are leaving theirs too'). This often comes up during debates about whether polluting sources, such as airports or heavy industrial plants, should be built or protected – supporters commonly argue that another country will build it if their country doesn't, as if the other country doesn't have its own climate commitments. Left unchallenged, this argument could make it impossible to win support for actions to reduce emissions. To avoid this global Tragedy of the Commons, there must be wider recognition that success depends on every country. If small countries don't cut their emissions, large ones couldn't be expected to do the same.

The perception that it is now too late to prevent climate change may be another source of inaction. This may be exacerbated by the argument, sometimes made by people who are worried about the problem, that the world is running out of time and that international agreements have failed to deal with rising emissions. The problem with this argument, if it is presented as simply as this, is that it gives the impression the issue is binary – the world either prevents climate change or it doesn't. But in fact the level of disruption the world will face is on a continuous scale. Humanity's actions will determine where the world falls on that scale, meaning emission cuts will almost always be worthwhile. Climate change has already started but the world can still stop dangerous global warming.

As well as showing that climate change isn't an 'all or nothing' problem, we will have a better chance of overcoming inertia if it is widely understood that the world is already making progress. If there is no hope of solving a problem, many people will be tempted to ignore it – what is the point in devoting effort to something that offers nothing but disaster? If climate change seems a story of despair, it is not surprising that many people don't want to pay attention to it. But when there is hope, everything is different. As humans, if there is a problem we can get stuck into, where we might be able to make a difference, we are more likely to be interested. This is reflected in the evidence about public attitudes to climate change. A 2010 study[9] found that people were more likely to approve of measures to cut emissions when the actions were described in terms of their potential to reduce damage, rather than in the slightly different terms of the damage that would occur if the world did nothing. The optimism trumped the despair.

So far, the focus of this chapter has been on countering pessimism about climate change, to show it doesn't need to be a Tragedy of the Commons, but we can do better than that. Not only can we show that significantly reducing emissions is possible, but we can also be optimistic about what that would mean for the world. This is controversial. There are strong arguments on both sides about how far we can take such optimism, as we will see later in this chapter.

Climate change is a menace that threatens to make life worse for billions of people. And yet research indicates there are benefits to talking about the positive side-effects of addressing climate change. An international study[10] found that describing the wider benefits of measures to cut emissions – beyond avoiding warming – appears to motivate action to address climate change, including among people who were unconvinced about the problem. The arguments that had the greatest impact were that cutting emissions would also produce economic, scientific and community benefits, and would make society more caring, moral and considerate.

Putting this together, the positive case for tackling climate change could be phrased as:

> 'It won't be easy to stop climate change – but the world is starting to deal with the threat and some of the changes we could make to reduce global warming would improve our communities in other ways. Nearly every country in the world – rich and poor, large and small – has agreed to limit greenhouse gases and, because of these efforts, global emissions have now stopped increasing, proving it is still possible to prevent dangerous climate change. But while it will be hard work to cut emissions further, it doesn't all need to involve sacrifices. The threat of climate change could motivate us to make changes that would improve our lives in other ways. Building, installing and running wind turbines and solar panels is already creating millions of new jobs; generating electricity from clean sources and switching to electric cars would make our air safer to breathe; and coming together to tackle a problem that threatens us all would help us build stronger communities.'

The bright side

In the months ahead of the 2015 Paris climate conference, close to every country in the world (between them covering 98.9 per cent of global emissions[11]) submitted statements detailing their planned contribution to cutting the world's emissions. These pledges were far from perfect. It was difficult to compare them, some were not really believable and the combined pledges were not enough to keep warming to safe levels.[12] But the subsequent agreement set a process for countries to strengthen their pledges and the concerted effort showed that the burden of cutting emissions was not just being left to a few rich countries. To take a few examples: China, the world's biggest emitter, declared a 2030 target of cutting by at least 60 per cent the quantity of emissions produced by each unit of economic activity. This could lead to its total emissions beginning to fall from the mid to late 2020s despite the country's predicted rapid growth[13] – a target that it looks on course to meet.[14] Brazil, the world's 12th-largest emitter, pledged to cut its emissions by 37 per cent by 2025.[15] Even Ethiopia, one of the poorest countries in the world, said it will cut its emissions by 64 per cent by 2030, compared with what they would have been if it took no such action.[16]

Of course, these are only intentions and if a country misses its target, international embarrassment may be the only punishment. But that is the nature of most international relations. The fact there is no real enforcement mechanism for these pledges isn't a trivial problem, but their voluntarism doesn't make them meaningless. The pledges are a signal of what each country is prepared to do if others also fulfil their own commitments. The main thing stopping a country from reneging on its pledge, other than the international shaming it might face, is the knowledge that abandoning emissions cuts could prompt other countries to do the same, until efforts to avoid disastrous global warming collapse. That should be enough to keep most governments committed, although it leaves the risk that a country whose government isn't worried about climate change or is even opposed to the principle of cutting emissions might decide that abandoning – or sabotaging – the Agreement would be worth the international criticism, as reflected by Donald Trump's announcement in June 2017 that he would take the US out of the Agreement.

And even if the Agreement survives on paper, perhaps we can't yet tell whether the biggest-emitting middle-income countries such as China and India are serious about their pledges. They might have no intention of holding up their end of the bargain. Maybe richer countries would be at a competitive disadvantage if they carried on cutting emissions on the assumption that China and India will do as they have promised. This is central to the arguments of some nationalists – like Trump – who claim that other countries are doing less to tackle climate change than their own country is.

That sounds a reasonable worry – and may have been a serious concern a few years ago – but the evidence now points in the opposite direction. Rapidly developing middle-income countries are making significant efforts to cut their emissions. China is expanding its wind- and solar-power capacity at an unprecedented speed and in 2015 it added more of both than any country has done before in a single year.[17] While its annual emissions nearly doubled between 2004 and 2014, that has now changed radically – in 2015 its emissions appear to have fallen slightly, and may have done so again the following year.[18]

So it is clear that efforts to tackle climate change are international and demonstrate that we do not need to give up all hope that the world will avoid dangerous warming. But how about the suggestion that tackling climate change can make our lives better in other ways?

Some of those worried about climate change criticize a particular kind of optimism about a low-carbon future, which they describe as 'bright-siding'.[19] A bright-sider is someone who argues we should have a positive outlook about climate change, not because they doubt the potential severity of its consequences, but because they believe humans can avoid disaster and, in the process, make the world a better place. Bright-siders believe a low-carbon future will be an appealing place to live in. So far, so mainstream. Why should this be unpopular with some climate campaigners? It sounds similar to some revolutionaries' vision of a future where emissions are cut in ways that make economic systems radically fairer.

There seem to be two reasons for the dislike of the bright-siders' vision of the low-carbon future. The first, where the bright-siders differ from some revolutionaries, is that they don't demand economic revolution. As George Marshall puts it, 'bright-siding is ultimately a regressive narrative that validates existing hierarchies... it promotes an aspirational high-consumption lifestyle while ignoring the deep inequalities, pollution and waste that make that lifestyle possible. And this is why, despite its upbeat tone, it is... unappealing to many ordinary people'.[20] The problem, for many critics, is that proponents of this emissions-cutting argument are comfortable working with the private sector, don't mind companies making a profit, and, ultimately, are okay with consumer culture.

It is certainly the case that economic growth has created both inequalities and the climate crisis, and that renouncing consumerism may make it easier to cut emissions – but that doesn't mean this criticism of bright-siding is fair. While ending growth might produce the deepest cuts in emissions, it may not be the only way to avoid calamitous warming, as discussed in the previous chapter. As long as we see climate change as a matter of degrees, rather than an all-or-nothing problem, it seems reasonable to consider how emissions could be cut within a growing economy. This matters because there doesn't seem to be much public appetite for relinquishing consumerism. The suggestion that 'aspirational high-consumption lifestyles' are unappealing to ordinary people reflects the tendency for many on the open Left to talk about climate change in terms that appeal to people like themselves, forgetting that many others – including the swings – have shown little sign of being ready to give up on consumption. Basing the argument for cutting emissions on a vision of the future that appeals to the Left is a certain way of making sure climate change continues to be seen as a leftwing problem.

The second criticism of bright-siding argues that the optimists are wrong to say the need to deal with climate change is good news. Critics suggest that the unwelcome effects of global warming so vastly outweigh the positive side that it is absurd to talk about it as anything other than a looming disaster.

Some measures to cut emissions would improve daily life, but they are rarely without downsides. One example is the potential for the growth of lower-emission technologies to create jobs. This certainly seems plausible – manufacturers of wind turbines, solar panels and batteries are booming. But at the same time, a coalminer or oil-field worker will not find such a positive outlook in emission cuts. Nicholas Stern, a British economist, calculated that the world can avert disastrous climate change with measures that would cost about one per cent of GDP each year, which would be far less than the damage caused by uncontrolled climate change[21] (he has since said that the cost-benefit balance is even further in favor of cutting emissions[22]). While his vision was a positive one, showing that dealing with the problem is affordable, he still concluded that cutting emissions would make the world worse off than it would be if climate change didn't exist.

And yet some of the potential benefits are enormous. One of the greatest is in cleaning up the world's air. Around 6.5 million people currently die each year as a result of air pollution.[23] This isn't just a problem in poor countries – annual air-pollution deaths in Australia are 3,300; in Canada, 7,700; in the UK, 28,300; and in the US, 99,000. Around half of global air-pollution deaths are the result of people breathing dangerous indoor air, mostly because of the use of coal, wood and animal dung for domestic cooking and heating, and this may not be directly affected by climate policies.[24] But even in countries most affected by this – as in richer places – coal-fired power stations and fumes from vehicles are also killers. The replacement of petrol or diesel vehicles with electric equivalents, and the closure of coal power plants, would bring enormous health benefits. Surprisingly, a study mentioned earlier in this chapter also found that the potential to reduce air pollution is fairly unlikely to motivate those people who aren't worried about global warming to support cuts to emissions.[25] Since the health benefits of clean air would be so great, the answer may be to put more work into showing the scale of the problem of air pollution and how it would be reduced with cleaner energy, rather than assuming that the swings already recognize this.

The description of the secondary benefits of tackling climate change finished with a reference to how it could make our communities stronger. This was another finding of that study – showing that tackling climate change will make people more considerate towards one another is an effective way of motivating action. But many people have never heard anyone saying this and so they may initially find it hard to believe. One way of making it more credible is through stories that show how local efforts to cut emissions and deal with the impacts of warming have brought communities together. For example, if local people are helped to own power-generating facilities, such as wind farms or solar plants, people in the community might get to know one another and feel they have control over how money is spent in their area.

With these examples in mind, is it fair to describe the need to deal with climate change as good news? On the face of it, critics of this approach seem right that there is a danger that we might unintentionally mislead people into thinking there is nothing to worry about. Unless you believe that the world will stop dangerous climate change by transforming economies in ways that fix other injustices, it is hard to see the need to cut emissions as, on balance, good news.

Yet there are grounds for optimism. The world certainly is capable of limiting climate change, at least to some extent. Most countries have begun to demonstrate their willingness to cut emissions; while the Paris pledges are still insufficient, their existence, and the recent slowdown in emissions growth, show that progress is possible. And some of the measures that would reduce emissions can indeed bring other benefits. Talking about these secondary benefits is likely to increase support for cutting carbon.

So it seems right to balance talk about the dangers of climate change with the good news about the world's ability to deal with it, and the benefits of doing so. We won't persuade the swings if we neglect either side – that avoiding dangerous global warming is an enormous challenge that will demand some tough decisions, but one the world is capable of achieving and which will produce some other benefits.

Sacrifices and solidarity

But this still leaves a problem. While it does indeed seem possible to talk about the positive side of some parts of dealing with climate change, we haven't yet confronted the really hard stuff. If our goal is to describe global warming in a way that doesn't utterly dishearten people, what does that mean for the challenges that no-one wants to face up to?

In my first 12 months of working on this book, I became a Silver Status frequent flier. To reduce my contribution to climate change, I cut back on my meat consumption, from eating it most days to having it just once every week or two, and I mostly stopped having milk on my morning cereal. That was hard at first, because I love a good burger, but it quickly felt normal and it reduced my personal emissions by the equivalent of nearly a ton of carbon dioxide a year.[26] Since the limit I should be aiming for is something like one ton a year[27] this seems a worthwhile contribution.

But it is far from enough. As long as I fly as often as I did in that year, my contribution to climate change is still way beyond what is sustainable. My carbon-dioxide emissions from flying alone were around three and a half tons.[28] Ouch. That's far more than my total emissions allowance, for everything in my life. If the world is to prevent dangerous warming, people like me are going to have to cut back on flying, or many other people will have to make even deeper cuts to compensate for my excesses. So long as there is no immediate prospect of a technological fix that will slash emissions from each flight, it is hard to see how we can continue to expand aviation while avoiding disastrous climate change. At a personal level, that is a depressing conclusion. I enjoy travel and couldn't reduce it as easily as I have my meat-eating. But something has to give.

Flying isn't the only problem like this – the world may have to cut back on other things, too. This will probably include meat and dairy consumption, as it is hard to see how its continued worldwide growth can be reconciled with bringing emissions down low enough that measures to absorb greenhouse gases can take overall emissions below zero. It may also mean consuming fewer manufactured goods, by making devices and appliances last longer and upgrading less often. The details will depend on the world's success in reducing the emissions from industrial processes and from generating electricity – if consumer goods can be made more cleanly, each new gadget won't release so much carbon. But regardless of the exact details, if the world is to avoid extreme climate change many people will have to make some such sacrifices.

There is no way to fudge this. While it is possible to talk about the benefits of some of the ways in which the world could cut emissions, there are limits to what most people would see as good news. When the future offers less freedom to take frequent foreign holidays, to eat meat twice a day, or to buy and throw away manufactured goods without much thought, most people – including many of the swings – would struggle to see the positive side. I am in the base segment, yet I still find it hard to accept that I should rein in my flying. The thought of not being able to travel nearly as much as I do now is both unwelcome and, as I don't want to it to be true, hard to believe fully at a deep level.

Winning acceptance and support for these sacrifices is ultimately what this book is about. If we are to build majority support for cutting emissions, we need to persuade the people who are apathetic about climate change that it really does matter, and that collective efforts are worthwhile because success is possible and outweighs the costs. Fortunately, I have found when talking about climate change that, by this point in the conversation, many people want to know what they personally can do. Often they are persuaded that there is a problem and they want to help – but they don't know how they can contribute. Many people are receptive to hearing about even these difficult challenges, as it gives tangible ways they can think about changing their own lives. But, even so, when it comes to the sacrifices that aren't easily compensated by a vision of a better world, not all the swings are readily persuaded that they should make such sacrifices: we are still missing something. To persuade more people to support the toughest measures, we also need political leadership.

This leadership needs to do a few things. The first is to make the case for why climate change matters and why richer countries need to take action. These are the arguments covered in Chapters 8 and 9, and they don't rely only on politicians. Anyone can show that climate change matters to their local community and that it is a problem everyone should care about – indeed, these arguments may be most effective when they are heard from family and friends. But political leaders have the ability to raise the profile of climate change more than anyone else. By talking about the problem far more often, they can turn it from something most people hear about only from non-profit campaigners, to an issue that is part of the national debate and becomes widely accepted as a top priority. And when we move beyond the arguments about the importance of climate change, to focus on the sacrifices that may be needed to address it, political leaders are essential.

Having to cut back on things we enjoy is never welcome news and politicians won't be taken seriously if they fail to acknowledge that. A politician would be laughed at for claiming that, for example, flying less is actually good news because it means we can better appreciate our own country. It would be seen as transparent spin. The only way for a politician to be taken seriously on the subject is for them to acknowledge that the need to fly less is unwelcome. It is so rare for a politician to openly give bad news that at least one who did this may be seen as serious and straight-talking.

But just saying the world will need to cut back on things it enjoys isn't enough; political leaders also need to show that the sacrifices will be fair. As we saw in Chapter 5, one of the strongest imperatives for the human mind is to avoid losses. It is inescapable that some measures to cut emissions will seem like losses – but at least they can be made to feel fair. Psychological research has also shown how fiercely most people resist situations where someone is deemed to be cheating or unjustly benefiting at the expense of others. For example, several studies have found that, when someone offers to split a pot of money in a way that seems grossly unfair, most people will reject the proposed division even if that means receiving nothing – behavior that would be irrational if people were only motivated by financial reward.[29] In a similar study, three-quarters of participants were so averse to unfairness that, in order to punish someone who had acted unfairly, not to themselves, but to someone else, they were prepared to receive less money. Nationalists who oppose measures to cut emissions understand this, which is why they put so much emphasis on the claim that their country is being expected to do more than others to deal with climate change. With this in mind, if we are to build support for sacrifices to cut emissions, we will have a much greater chance of success if the burden is seen to be fairly distributed.

Let's consider how this could work for aviation within a country. A traditional economists' way to restrict growth in passenger numbers may be to increase the price of tickets until demand falls to the appropriate level. This would work from an emissions perspective but it would worsen inequality in access to international travel. Only richer people would be able to fly – even more so than at the moment – so poorer people would bear the cost of avoiding dangerous global warming by sacrificing the ability to travel to far-off places. An alternative, which some people campaign for, would be to provide allowances for flights so everyone could fly, perhaps once a year, at a lower price if they wanted to and could afford it, with subsequent flights being more expensive.[30] This level of government interference may be too heavy-handed to win widespread approval easily but, compared with the less well-off having to shoulder the burden of cutting emissions from flying, it may be more appealing than the alternatives.

To be seen as fair, any sacrifices also need to feel like a country's own choice, rather than having been imposed by some external force. It was by tapping into anger against supposed distant rulers in Brussels, with the slogan 'Take back control', that campaigners won the 2016 referendum to take the UK out of the European Union. Support for sacrifices to cut emissions will never be wide enough if they are commonly seen to be a diktat from outsiders.

As with the other most difficult parts of dealing with climate change, this is only going to happen if political leaders acknowledge the problem and show that, in tackling it, we are all in it together. It is common for people worried about climate change to invoke memories of the sacrifices shown in the Second World War as an example of what a society can achieve if it pulls together. That comparison is often overdone – until so much carbon has been released that the world is already committed to dangerous warming, the impacts of climate change won't be as dramatic in most places as a total war. But when it comes to some of the sacrifices that may be needed to cut emissions, the wartime comparisons are relevant. They show that people will cut back on some things they enjoy for the sake of a common purpose. In short, they show societies can act out of solidarity.

This solidarity is essential for the hardest aspects of cutting emissions. A person naturally has the most solidarity for those they already care for. But, once we have strengthened swings' interest in climate change, the same arguments can be extended to people in other countries and those yet to be born. We can show that avoiding extreme climate change is a moral imperative, not just another policy question that can be given more or less priority according to the mood of the day.

This might seem a long way from the suggestion that we should appeal to self-interest and focus on how climate change will affect people in richer countries, but in fact it is the final stage of a process that began with other arguments. Currently, around half of the population of some high-emitting countries are apathetic about climate change. They broadly accept it is real, but they just aren't that interested in the problem. Compared with other issues, global warming doesn't seem particularly urgent to them. This leaves efforts to cut emissions vulnerable to arguments that international co-operation is not worthwhile and countries should instead prioritize what nationalists present as their country's self-interest. Before talking about the solutions to climate change, we have more work to do to set out the problem and explain why it is wrong to suggest that it doesn't pose a threat to rich countries. When we jump straight to telling the swings that they should cut back on flights and meat, we are met with incredulity. Likewise, before we can expect more people to consider climate change a matter for international solidarity, they need to be persuaded that the effort is worthwhile. That depends on them accepting that climate change matters to them and that emissions cuts can still avert disaster. It is only after those arguments have been won that we can hope for a receptive audience for appeals on behalf of people in the future and in far-off places.

Hope is the bridge that takes us from talking about the problem to talking about the solutions. It allows us to show that the world is capable of preventing dangerous warming and that doing so will bring wider benefits. In the conversation about solutions – with the prospect that these may be within reach – we should be frank about the sacrifices that might be needed to cut emissions. In some of these sacrifices there is not much to celebrate but, compared with the threat the world faces, even those are relatively minor. An honest conversation about where we are and what we can achieve can use hope to foster solidarity and give the world a better chance of avoiding disaster.

Conclusion: In the balance

The 2015 Paris climate conference offered a glimpse of a world where governments make serious commitments to prevent disastrous climate change. In the final deal, leaders agreed to keep warming within safe limits and spelled out what their countries would do to begin to achieve that.

This Paris Agreement was reached without an overwhelming public clamor for action. On the day before the conference opened, around 600,000 people around the world took part in demonstrations. It was reportedly the largest climate protest ever,[31] and yet it represented less than one ten-thousandth of the world's population. Earlier that year, nearly twice as many people had signed a petition calling for the British TV presenter Jeremy Clarkson to be reinstated after he was fired for assaulting a producer.[32]

So perhaps the swings don't matter. Climate apathy wasn't confronted and governments still signed up to an ambitious deal.

But let's take another view. In the six months before the Paris conference, the UK government had systematically undermined its own climate policy. Soon after its May 2015 election victory, the Conservative government announced it would abandon subsidies for solar power and onshore wind turbines, remove tax incentives for electric cars and community-energy schemes, and cut support for carbon capture and storage.[33] It did so despite knowing that it was already on course to miss its legal emissions target in the mid-2020s.[34] While there is widespread public support in the UK for the goal of cutting emissions, these policy reversals prompted little outcry beyond those who care deeply about global warming. And this is a government that says it considers climate change to be a serious threat. Less than a year after the Paris conference, the US elected a president who had said global warming was a Chinese hoax and went on to announce he would pull out of the

Agreement altogether. If the US does indeed follow through on Trump's declaration and renounces its climate commitments other countries may be tempted to do the same.

As long as concern about climate change is confined to a small portion of the population, the difficult emission-cutting measures will be dispensable. Widespread public agreement that climate change deserves attention means that most governments think there is an electoral advantage to being seen to be dealing with it. This helps explain why so many leaders were willing to sign the Paris Agreement – and why those who openly dismiss the threat of climate change are rare. But setting targets is easier than achieving them. When meeting the targets means forcing voters to accept costs in their everyday lives – through higher bills to fund clean energy, increased taxes on more-polluting vehicles or restrictions on flying or meat consumption – politicians will put off the hard decisions. If only a small proportion care passionately about tackling climate change, the cost to politicians of making decisions that eventually lead to missed emissions targets – which will probably only be missed after they have left office – will be less than the cost of the ire they may face from voters who have been forced to accept sacrifices. The recent wave of nationalism may surge further and incline the world towards ever more isolationism, or it may relent. But even if nationalism declines, it is hard to see the world doing enough to avoid disaster while so many people remain apathetic.

This seems a daunting barrier. What are the prospects that it can be overcome?

Recent experience isn't encouraging. In the past few years, there has been little sign of any growth in public concern. People in richer high-emitting countries are, on average, less worried about climate change now than they were over a decade ago. Warnings about the looming crisis from scientific, religious and community leaders seem to have done no more than avoid any further deterioration in public concern.

We might choose to wait for the effects of a warming climate to become sufficiently visible to jolt the world into radical action to avoid worse disaster. This is possible, but recent experience is again discouraging. The weather extremes of the past few years don't seem to have had a sustained impact on public opinion. When they have prompted more worries about global warming, those increased fears have been just a temporary spike that has quickly subsided. So far these extremes don't seem to unite public opinion behind meaningful action.

So if we can't expect climate apathy to disappear in the face of reality, we are left with the question of how we can hasten its demise. This book has set out some of the answers to this, focusing on what we can do to influence the (roughly) 50 per cent of the population who accept that climate change is real and a problem, but don't pay much attention to it and aren't yet willing to make sacrifices in the name of cutting emissions.

This apathy is partly a result of how most people think about the threat that climate change poses, which is shaped by how those of us worried about the problem talk about it. Currently, the consequences of global warming are often described in terms that make it seem distant, slow-moving and undramatic. We use small numbers to talk about average planetary warming and annual sea-level rise, or warn about polar bears' habitat, when we would have more chance of success if we focused on what uncontrolled global warming would mean for the people we are talking to, with examples like storms, floods, ferocious heat waves, wildfires and droughts.

Arguments about climate change have become a battle between the open Left and the free-market Right, allowing the people who want to delay action to give the impression that those arguing for emission cuts are disingenuously doing so for other reasons. Onlookers see a polarized battle and assume that the problem has been exaggerated, with the truth somewhere between the extremes. Instead of allowing this perception to continue, we can show that worries about the issue are spread across the political spectrum, by focusing on the controversies about *how* we tackle climate change – not *whether* we should tackle it.

Climate change can seem both depressing and inevitable and most people don't want to pay attention when there appears to be no hope of a positive outcome. This, too, we can do something about. Without ever claiming that climate change is good news – it isn't – it is possible to show that avoiding disastrous warming is still within the world's reach and that measures to do so can be fair and can also make the world better in other ways.

So, what next? If you are persuaded that climate apathy is indeed a problem, and that changing the debate has the potential to help the world avoid disaster, there are several things you can do.

First, don't give up on activism. Despite the risks of climate change being too closely associated with the open Left, it is clear that protest and direct action produce results. From brave stunts, such as blockading the transport of oil rigs, to mainstream actions such as customer boycotts, marches and phone calls to politicians, activism puts pressure on governments and businesses to take climate change seriously. Such direct activism may well become more important in places where past achievements come under threat, as may be the case during Trump's presidency.

There is also a particular role for anyone who has influence in organizations that can reach the swing audiences. There are already campaign groups that mobilize people in the base segment. But so long as those organizations need to continue to serve their members and supporters, they can't easily reach people in the swing groups at the same time. Yet there are many other large and respected organizations – charities, not-for-profits and businesses – that are quietly worried about climate change and are trusted by people in the swing segments. At the moment, they don't say much about the issue, leaving it to the high-profile campaigners to be the most prominent. With the swings so important to the future of the climate, these organizations have a moral responsibility to use their voice to show that global warming isn't just an issue that interests environmentalists on the open Left – it is something that matters to them and their audiences as well.

And, most importantly, everyone can tackle apathy by talking about climate change with those among their friends and family who are unsure about the problem. You may not be a runway-blocking climate activist, but if you are worried about the problem and want stronger action to deal with it, this is what you can do, right now. If every one of us who is in the base segment identifies three or four people in the swing segments and changes their minds, we will transform the world. There are currently more than twice as many swings as there are people in the base segment. If each person who is worried about climate change succeeded in winning over just one or two of the people that aren't so worried, those figures would be reversed.

How can those conversations persuade swing voters? It is partly about the evidence and examples – the previous chapters have described what is most likely to interest the swings. But it is also about the messenger. Most people say they don't particularly trust what their friends and family say about climate change. To overcome that, anyone hoping to convert some swings needs to become recognized as knowledgeable. This takes effort, but it isn't hard or expensive as there are many excellent climate websites that provide easy-to-follow explanations of the important issues and which don't require prior knowledge. I recommend Carbon Brief (carbonbrief.org) and Vox (vox.com) for balanced, accessible and informative explanations of current debates about how the world can cut emissions and what will happen if it doesn't. Skeptical Science (skepticalscience.com) is good for providing evidence to rebut the arguments used to justify inaction. Climate Home (climatechangenews.com) is my favorite for keeping up to date with international climate-change news. Some of these sites provide free daily and weekly summary emails. Spending 10 minutes a day reading these is a small investment in understanding much better the arguments that fly around in the debate and in being able to talk credibly about it.

Such conversations about climate change will be most effective if they take those of us in the base segment out of our comfort zones. It isn't enough for us to talk only to those who share our values about how climate change deserves their attention – we also need to win over those who think differently and aren't motivated by the same vision of the future. There are people who want to tackle climate change but who have different political outlooks, and we should help them be heard in the debate. While we may profoundly disagree with them on other issues, we are working to the same ends when it comes to cutting emissions, and their visible participation in the debate can broaden its appeal among those who are currently unpersuaded.

Would this be enough? If the debate shifted, so more people recognize how extreme climate change would affect the people and places that matter most to them; that it is more than a Left-Right political argument; and that the world can deal with it in ways that make our lives better, would climate apathy fall? The truth is, I don't know. But the arguments I have suggested to influence the swings are backed up by the evidence that is available. If we want to build a climate majority to back the measures that will be needed to avoid disaster, we have to find ways of persuading those who have, so far, remained outside the debate.

References

Introduction

1 Climate change or global warming? Both are accurate in relation to the consequences of recent and future human activities – the warming of the globe caused by an increase in greenhouse-gas concentrations will change the climate. Some studies have shown that the term global warming is more likely to elicit concern among the public – particularly in the US – so perhaps that should be our preferred term if our goal is to persuade more people it matters. Yet this doesn't seem to be true everywhere and among people who work on the problem, climate change has become the preferred term to the extent that it now seems odd in some circles to talk about global warming. But, to avoid repetition, I often refer to global warming and mostly use the terms interchangeably. **2** Reuters, nin.tl/bayer-apology **3** Obviously this is more true of democracies and other countries where governments are sensitive to public opinion. The reality of electoral systems also means that politicians often depend on the support of particular voters and may consider others to be unwinnable, meaning some voters are, in electoral terms, more important than others. But for the purposes of this book, it is enough that democratic governments are generally influenced by public opinion. **4** I first came across the political model when I worked at the research agency, PSB, which used the approach in its work with private-sector and political campaigns. **5** The only other option would be to hold out hope of an engineering solution, either to remove emissions from the air or to reduce the amount of sunlight reaching us. As discussed in Chapter 3, the former of these could provide only a small amount of extra time to cut emissions, while the latter might be better than disastrous climate change but would bring new risks that make it an unattractive option so long as it is still possible to cut emissions. **6** For example, *Mother Jones*, nin.tl/strange-relationship **7** This excludes emissions from land use, as the annual variability of these makes it difficult to judge progress. Jos GJ Olivier, Greet Janssens-Maenhout, Marilena Muntean & Jeroen AHW Peters, *Trends in global CO2 emissions: 2016 Report*, PBL Netherlands Environmental Assessment Agency, European Commission, Joint Research Centre (JRC), The Hague, 2016, nin.tl/emission-trends-2016 **8** For example, Oliver Morton, *The Planet Remade: How Geoengineering Could Change the World*, Granta Books, 2015; George Marshall, *Don't Even Think About It: Why Our Brains Are Wired To Ignore Climate Change*, Bloomsbury, New York, 2014.

Chapter 1 – The wrong target

1 Sun Tzu, *The Art of War*. **2** Ziva Kunda, 'The case for motivated reasoning', *Psychological Bulletin*, November 1990. **3** ABC, nin.tl/impartiality-guidance; BBC, nin.tl/ed-guidelines; CBC, nin.tl/journo-standards **4** BBC, nin.tl/climate-change-factor **5** BBC, nin.tl/complaints-unit-finding **6** BBC, nin.tl/science-coverage-review **7** For example, Carbon Brief Staff, nin.tl/response-to-ridley **8** *Independent Australia*, nin.tl/denial-easy-abc **9** iPolitics, nin.tl/follow-bbc **10** Media Matters for America, nin.tl/media-sow-doubt; *Carbon Brief*, nin.tl/media-coverage-graphs **11** Nelya Koteyko, Rusi Jaspal & Brigitte Nerlich, 'Climate change and 'climategate' in online reader comments: a mixed methods study', *The Geographical Journal*, March 2013. **12** *The Sydney Morning Herald*, nin.tl/climate-bromance **13** Executive Office of the President, 'The President's Climate Action Plan', The White House, June 2013. **14** Adaptation Sub-Committee, 'How well prepared is the UK for climate change?', Committee on Climate Change Adaptation, September 2010. **15** Pope Francis, 'Laudato Si'', The Vatican, 24 May 2015. **16** John Cook et al, 'Quantifying the consensus on anthropogenic global warming in the scientific literature', Environmental Research Letters, May 2013. **17** IPCC, *Climate Change 2014: Impacts, Adaptation, and Vulnerability. Part A*, Cambridge University Press, 2014. **18** Patrick J. Buchanan Official Website, nin.tl/buchanan-speech **19** Pew Research Center, nin.tl/party-id **20** *The Spectator*, nin.tl/tory-away-day **21** Or whoever they are addressing – the same applies to any collection of people or entities, from parliamentarians to businesses. **22** Chris Rose, *What Makes People Tick: The Three Hidden Worlds of Settlers, Prospectors and Pioneers*, Matador, 2011. **23** These segments are similar to, but not quite the same as, those often mentioned in discussions about political strategy, like the US Soccer Moms and the UK Mondeo Man. A difference is that those segments usually start with strategists' views of the required strategy ('we need to reach lower-middle-class voters with children') and a notional segment is identified to make it easier to describe those voters. Equally, political campaigns often do identify target groups based on their demographics, rather than on the basis of more sophisticated segmentations. This is mostly about money – it is easier to identify someone on the basis of their wealth, age, ethnicity and family status than it is to identify them based on what they think, though the former approach is usually less useful. **24** Gillian Tett, 'Donald Trump's campaign shifted odds by making big data personal', *Financial Times*, 26 January 2017. **25** In the private sector they are often labelled 'reach', a euphemism making the point that you would have to reach quite far to win them over. **26** I once worked with an Italian who always, in apparent innocence, referred to these segments as 'swingers'. I'm pretty sure he knew just what that meant! **27** But not too many to be useful. The purpose is to create a model that simplifies the real world so campaigns can plan how they will target different audiences. As a model identifies more groups, it becomes more accurate, but less useful as a basis for campaigners to address as many people as they need to. This is why well-resourced campaigns that can micro-target tiny segments, can now use segmentations particularly effectively. **28** Pew Research Center, nin.tl/pew-climate-study **29** Ipsos MORI, nin.tl/cardiff-uni-study **30** The Lowy Institute, nin.tl/lowy-climate-poll **31** Gallup, nin.tl/gallup-us-concern **32** Stuart Capstick et al, 'International trends in public perceptions of climate change over the past quarter century', WIREs Climate Change, January/February 2015. **33** Nielsen, nin.tl/aus-nielsen-report; Angus Reid Public Opinion, nin.tl/belief-in-bigfoot **34** YouGov, nin.tl/diana-conspiracy **35** Rasmussen Reports, nin.tl/conspiracies-abound **36** A copy of the memo is available at nin.tl/luntz-research **37** Naomi Oreskes & Erik M Conway, *Merchants of Doubt: How a Handful of Scientists Obscured the Truth on Issues from Tobacco Smoke to Global Warming*, Bloomsbury, 2010. **38** For example, *DeSmog*, nin.tl/funders-dark-money **39** Amelia Sharman, 'Mapping the climate sceptical blogosphere', *Global Environmental Change*, May 2014.

Chapter 2 – Who cares about climate change?

1 Michael Porter, 'What is Strategy?', *Harvard Business Review,* Vol 74, No 6, 1996. **2** In fact the accord was the basis for individual countries making commitments to cut their emissions, which in turn was the basis for the 2015 Paris Agreement, so in retrospect it wasn't as much of a disaster as it seemed at the time. **3** The original research is no longer available online, but a summary of it, which includes the finding that 64 per cent of those who think climate change is natural were dissatisfied that the Accord was non-binding, is available at: nin.tl/satisfied-brits **4** Not all swings think climate change is a natural phenomenon – as we will see, many are more convinced that it is happening, caused by humans and likely to be a problem. **5** For example, CSIRO, nin.tl/aus-climate-attitudes; Université de Montréal, nin.tl/canada-climate-attitudes; Carbon Brief Climate Poll, nin.tl/uk-climate-attitudes; EES, nin.tl/us-climate-attitudes **6** For example, YouGov, nin.tl/intl-survey **7** Pew Research Center, nin.tl/public-support-climate-action **8** For example, Ipsos MORI, *Issues Index: 2007 onwards*, 2017, nin.tl/important-issues-uk **9** Carbon Brief, nin.tl/floods-change-debate **10** *The Scientist*, nin.tl/life-as-target **11** For example, Noise of the Crowd, nin.tl/opinion-changes **12** Anthony A Leiserowitz et al 'Climategate, Public Opinion, and the Loss of Trust', *American Behavioral Scientist*, 2013. **13** Opinium Wave 2, 2013a, nin.tl/carbon-brief-poll **14** CBC News, 'Canadians lack trust in some scientists, poll suggests', nin.tl/canadians-lack-trust (despite the article title, the poll found 65 per cent trust climate scientists, compared with 28 per cent who do not); YouGov, nin.tl/american-trust-in-scientists **15** For example: Ipsos MORI, nin.tl/trust-in-professions; Gordon Gauchat, 'Politicization of Science in the Public Sphere: A Study of Public Trust in the United States, 1974 to 2010', *American Sociological Review*, April 2012. **16** Anthony A Leiserowitz et al, 2013, op cit. **17** During part of the time I was writing this book I was employed as a consultant by the European Climate Foundation, which funds Carbon Brief. **18** Fake in the sense that we made them up, rather than that these stories were untrue and all the others were accurate. One of our silliest ideas for a fake story, that climate change is making horses smaller, turned out to already have been covered, apparently in all seriousness, by a well-known British news site. **19** Opinium, Carbon Brief Climate Poll, Wave 2, 2013, nin.tl/global-warming-tables **20** Anthony A Leiserowitz, *Public Understanding of Climate Science and Trust in Scientists*, nin.tl/yale-trust-in-scientists **21** *New York Times*, nin.tl/nyt-trump-interview **22** Stephan Lewandowsky, James S Risbey & Naomi Oreskes. 'On the definition and identifiability of the alleged "hiatus" in global warming', *Nature Scientific Reports*, 5, November 2015. **23** Unpublished UK online poll, September 2014. **24** I used the calculator created for the 2013 Warsaw climate conference, available at: nin.tl/climate-calculator. I set all inputs to the mid-level, changing only the ones I was testing, and assumed two long-haul return flights a year, of eight hours each way. **25** Chatham House, December 2014, nin.tl/livestock-effect **26** Canada.com, nin.tl/canadians-dont-know **27** *The Guardian*, nin.tl/power-subsidies **28** This quote is drawn from my own research, which is described later in this chapter. **29** Private interview. **30** There are other segmentation studies focusing on attitudes to climate and the environment. I focus on these as they are the most similar, so their resulting segments can be treated as equivalent. No climate segmentations of the population of Canada seem to have been published. **31** National Climate Change Adaptation Research Facility, nin.tl/enhancing-comms **32** Anthony Leiserowitz, Edward Maibach & Connie Roser-Renouf, 'Global Warming's Six Americas: An Audience Segmentation Analysis', nin.tl/six-americas; segment sizes are from: Yale University and George Mason University, nin.tl/six-americas-and-election **33** Defra, nin.tl/environmental-framework **34** Three Worlds, nin.tl/beyond-class **35** Defra, nin.tl/environmental-framework **36** CSIRO, nin.tl/climate-opinion-aus **37** Ibid. **38** The people I spoke to weren't representative of the swing segments, so I didn't attempt to use their responses to challenge the results of the academic research. The purpose of speaking to them was to help me understand the motivations for the attitudes that the other research had identified. **39** They are named *Uncertain* in the Australian study, but the label *Cautious* seems better to encapsulate what makes them different from the other swing groups.

Chapter 3 – Maps and roadblocks

1 David JC MacKay, *Sustainable Energy – without the hot air*, UIT, Cambridge, 2008. **2** BBC, nin.tl/chemicals-legislation **3** Short for Registration, Evaluation and Authorization of Chemicals. **4** BBC, nin.tl/chemicals-legislation **5** Business Green, nin.tl/manu-unaware **6** There was some public campaigning on REACH, for example by WWF. **7** This level is usually expressed in terms of the average planetary warming. But as Chapter 6 discusses, descriptions of climate change in terms of 'small number' global average temperature rises like 1.5°C are misleadingly reassuring for most people. For that reason I am avoiding using such figures. **8** Oxfam International, nin.tl/survival-fittest **9** International Institute for Environment and Development, nin.tl/iied-adaptation **10** This does not include land use changes, which vary quite widely from year to year so are generally excluded from calculations of the overall trend. Corinne Le Quéré et al, 'Global Carbon Budget 2016', in *Earth System Science Data*, November 2016. **11** Robert Mendelsohn, Ariel Dinar & Larry Williams. 'The distributional impact of climate change on rich and poor countries', in *Environment and Development Economics*, April 2006. **12** UNEP, nin.tl/adaptation-gap **13** Joeri Rogelj, 'Energy system transformations for limiting end-of-century warming to below 1.5°C', in *Nature Climate Change*, May 2015. **14** Based on Jos GJ Olivier, Greet Janssens-Maenhout, Marilena Muntean & Jeroen AHW Peters, *Trends in global CO2 emissions: 2016 Report*, PBL Netherlands Environmental Assessment Agency, European Commission, Joint Research Centre (JRC), The Hague, 2016, nin.tl/emission-trends-2016, and Joeri Rogelj, ibid. **15** For simplicity, these data are for the US only (chosen as it is the largest emitter among richer countries). Each country has a distinct profile of emissions; for example, nearly twice the proportion of Australia's emissions are from agriculture, but the overall picture is similar. US Environmental Protection Agency, nin.tl/US-greenhouse (see Table ES-7). **16** Don't forget that this is about the US: other countries have lower road emissions. **17** ICCT, nin.tl/real-world-fuel **18** Department for Business, Innovation and Skills, nin.tl/con-ind-influence **19** Kenneth Gillingham, David Rapson & Gernot Wagner, 'The Rebound Effect and Energy Efficiency Policy', RFF Discussion Paper, nin.tl/rebound-energy **20** William J Ripple et al, 'Ruminants, climate change and climate policy', *Nature Climate Change*, December 2013. **21** Reuters, nin.tl/jump-ship **22** Halvor Mehlum, Karl Moene & Ragnar Torvik, 'Institutions and the Resource Curse', *The Economic Journal*, January 2006. **23** For example, *American Scientist*, nin.tl/smil-article **24** Tyndall Centre for Climate Change Research Working Paper, nin.tl/BECC-review **25** For a clear description of these technologies and their potential, see Tim Flannery, *Atmosphere of Hope: Solutions to the Climate Crisis*, Penguin, 2015. **26** Joeri Rogelj, 2015, op cit. **27** Figure 1a in Joeri Rogelj, 2015, op cit. **28** Personal communication with Joeri Rogelj. **29** Corinne Le Quéré et al, 2016, op cit. **30** Compared with RCP 8.5, Keywan Riahi, Shilpa Rao, Volker Krey et al, 'RCP 8.5 – A scenario of comparatively high greenhouse gas emissions', *Climatic Change*, November 2011. **31** Oliver Morton comes to a more optimistic conclusion in his review of these options: Oliver Morton, *The Planet Remade: How Geoengineering Could Change the World*, Granta Books, 2015. **32** BBC, nin.tl/flood-money **33** Department of Energy & Climate Change, nin.tl/2015-energy-consumption (see Chart 5). **34** *The Guardian*, nin.tl/toaster-plans-dropped **35** While others have reached the same conclusion, the energy writer David Roberts was the first person I saw describing it in such clear terms. **36** Greenpeace & Sierra Club, nin.tl/final-boom-bust **37** For example, Opinium, 2013b, op cit. **38** BBC, nin.tl/impossible-house **39** For example, UK: Survation, nin.tl/october-issues; Canada and US: Canada 2020, nin.tl/comparative-climate **40** BBC, nin.tl/wood-fuel-plan **41** Carbon Brief, nin.tl/aviation-analysis **42** ICAO, nin.tl/environmental-protection **43** Supplementary Figure 2a in Joeri Rogelj, 2015, op cit. **44** Marco Springmann, H Charles J Godfray, Mike Rayner, & Peter Scarborough, 'Analysis and valuation of the health and climate change cobenefits of dietary change', *PNAS*, April 2016. **45** European Commission, nin.tl/shipping-emissions. These emissions may be higher than those estimates, for example: *The Guardian*, nin.tl/scale-of-emissions **46** OECD, nin.tl/meat-data

Chapter 4 – The stakes

1 Jared Diamond, *Collapse: How Societies Choose to Fail or Survive*, Viking Penguin, 2005. **2** Polling evidence on this is limited, but what there is backs up the point. For example, Opinium, 2013a, op cit. **3** Sayers and Partners, nin.tl/future-flood-risk, p112. **4** £615 million ($755 million) out of £732 billion ($898 billion) in 2014-15. HM Treasury, nin.tl/uk-budget-2014; Sara Priestley & Tom Rutherford, 'Flood risk management and funding', House of Commons Library, nin.tl/flood-risk-management. **5** Committee on Climate Change, nin.tl/risk-2017 **6** Carbon Brief, nin.tl/good-news-for-wine **7** John Porter, Liyong Xie et al, 'Food security and food production systems', in Christopher B Field, Vicente R Barros et al, *Climate Change 2014: Impacts, Adaptation, and Vulnerability. Part A: Global and Sectoral Aspects. Contribution of Working Group II to the Fifth Assessment Report of the Intergovernmental Panel on Climate Change*, Cambridge University Press, 2014. **8** Katharine Hayhoe, Scott Sheridan, Laurence Kalkstein & Scott Greene, 'Climate change, heat waves, and mortality projections for Chicago', *Journal of Great Lakes Research*, Vol 36, No 2, 2010. **9** Ibid. **10** BBC, nin.tl/syria-conflict-story **11** Eurostat, nin.tl/asylum-stats **12** Ricardo M Trigo, Célia M Gouveia, David Barriopedro, 'The intense 2007-2009 drought in the Fertile Crescent: Impacts and associated atmospheric circulation', *Agricultural and Forest Meteorology*, August 2010. **13** Colin P Kelley, Shahrzad Mohtadi, Mark A Cane, Richard Seager & Yochanan Kushnir, 'Climate change in the Fertile Crescent and implications of the recent Syrian drought', *PNAS*, March 2015. **14** Troy Sternberg, 'Chinese Drought, Wheat, and the Egyptian Uprising: How a Localized Hazard Became Globalized', in Caitlin E Werrell & Francesco Femia (eds), 'The Arab Spring and Climate Change: A Climate and Security Correlations Series', *Centre for American Progress*, February 2013, nin.tl/climate-arab-spring **15** Ibid, and Sarah Johnstone and Jeffrey Mazo, 'Global Warming and the Arab Spring', in Caitlin E Werrell & Francesco Femia (eds), 2013, op cit. **16** Syria: Colin P Kelley et al, 2015, op cit. Russia: Friederike Otto, Neil Massey, Geert Jan van Oldenborgh, Richard Jones & Myles Allen, 'Reconciling two approaches to attribution of the 2010 Russian heat wave', *Geophysical Research Letters*, February 2012. **17** Cynthia Rosenzweig & William Solecki, 'Hurricane Sandy and adaptation pathways in New York: Lessons from a first-responder city', *Global Environmental Change*, September 2014. **18** Madeleine Lopeman, George Deodatis & Guillermo Franco, 'Extreme storm surge hazard estimation in lower Manhattan', *Natural Hazards*, March 2015. **19** Radley Horton, Christopher Little, Vivien Gornitz, Daniel Bader & Michael Oppenheimer, 'New York City Panel on Climate Change 2015 Report, Chapter 2: Sea Level Rise and Coastal Storms', *Annals of the New York Academy of Sciences*, 2015. **20** Ibid. **21** Claudia Tebaldi, Benjamin H Strauss & Chris E Zervas, 'Modelling sea level rise impacts on storm surges along US coasts', *Environmental Research Letters*, March 2012. **22** Nova Scotia, nin.tl/coastal-tech-report, Chapter 7. **23** Climate Council of Australia, nin.tl/hungry-nation **24** Afshin Ghahramani & Andrew D Moore, 'Climate change and broadacre livestock production across southern Australia. 2. Adaptation options via grassland management', *Crop & Pasture Science*, August 2013. **25** Climate Council of Australia, 2015, op cit. **26** Ben Kirtman, Scott B Power et al, 'Near-term Climate Change: Projections and Predictability', in Thomas F Stocker & Dahe Qin et al, *Climate Change 2013: The Physical Science Basis. Contribution of Working Group I to the Fifth Assessment Report of the Intergovernmental Panel on Climate Change*, Cambridge University Press, 2013, Figure 11.19; Jelle Bijma, Hans-O Pörtner, Chris Yesson & Alex D Rogers, 'Climate change and the oceans – what does the future hold?', *Marine Pollution Bulletin*, September 2013. **27** Bärbel Hönisch et al, The Geological Record of Ocean Acidification, *Science*, March 2012. **28** Mebrahtu Ateweberhan, David A Feary, Shashank Keshavmurthy, Allen Chen, Michael H Schleyer & Charles RC Sheppard, 'Climate change impacts on coral reefs: Synergies with local effects, possibilities for acclimation, and management implications', *Marine Pollution Bulletin*, September 2013. **29** Katja Frieler et al, 'Limiting global warming to 2°C is unlikely to save most coral reefs', *Nature Climate Change*, September 2012. **30** Maria Byrne, 'Impact of ocean warming and ocean acidification on marine invertebrate life history stages: Vulnerabilities and potential for persistence in a changing ocean', *Oceanography and Marine Biology: An Annual Review*, Vol 49, 2011. **31** Camilo Mora et al, 'Biotic and Human Vulnerability to Projected Changes in Ocean Biogeochemistry over the 21st Century', *PLOS*, October 2013. **32** CSIRO and Bureau of Meteorology, nin.tl/climate-change-australia **33** Ibid. **34** Ministry for the Environment, nin.tl/nz-climate-predictions **35** Adaptation Sub-Committee, nin.tl/climate-change-scenarios **36** *CCRA2: Updated projections for water availability for the UK*, nin.tl/updated-projections-uk **37** Ministry for the Environment, 2016, op cit. **38** Gabriele Villarini & Gabriel A Vechhi, 'Projected Increases in North Atlantic Tropical Cyclone Intensity from CMIP5 Models', American Meteorological Society, May 2013. **39** Benjamin I Cook, Toby R Ault & Jason E Smerdon, 'Unprecedented 21st century drought risk in the American Southwest and Central Plains', *Science Advances*, Feb 2015. **40** For a summary of research on drought during the Dust Bowl and in projections for the late

21st century, see: Think Progress, nin.tl/hell-and-high-water **41** Katharine Hayhoe et al, 2009, op cit. **42** MD Flannigan, KA Logan, BD Amiro, WR Skinner & BJ Stocks, 'Future area burned in Canada', *Climatic Change*, September 2005. **43** John A Church, Peter U Clark et al, 'Sea Level Change', in Thomas F Stocker & Dahe Qin et al, 2013, op cit. **44** Ibid. **45** Climate Central, nin.tl/mapping-choices **46** Ian Joughin, Benjamin E Smith & Brooke Medley, 'Marine Ice Sheet Collapse Potentially Under Way for the Thwaites Glacier Basin, West Antarctica', *Science*, May 2014. **47** Johannes Feldmann & Anders Levermann, 'Collapse of the West Antarctic Ice Sheet after local destabilization of the Amundsen Basin', *PNAS*, November 2015. **48** Climate Central, nin.tl/mapping-choices **49** Kurt M Cuffey & Shawn J Marshall, 'Substantial contribution to sea-level rise during the last interglacial from the Greenland ice sheet', *Nature*, April 2000. **50** C Tarnocai, JG Canadell, EAG Schuur, P Kuhry, G Mazhitova & S Zimov, 'Soil organic carbon pools in the northern circumpolar permafrost region', Global Biogeochemal Cycles, June 2009. **51** Note that there is 1 ton of carbon in 3.67 tons of carbon dioxide. **52** Anton Vaks et al, 'Speleothems Reveal 500,000-Year History of Siberian Permafrost', *Science*, April 2013. **53** Valérie Masson-Delmotte, Michael Schulz et al, 'Information from Paleoclimate Archives', in Thomas F Stocker & Dahe Qin et al, 2013, op cit. **54** Yadvinder Malhi et al, 'Exploring the likelihood and mechanism of a climate-change-induced dieback of the Amazon rainforest', *PNAS*, December 2009. **55** Sybren Drijfhout et al, 'Catalogue of abrupt shifts in Intergovernmental Panel on Climate Change climate models', *PNAS*, October 2015.

Chapter 5 – Sight and mind

1 Center for Science and Technology Policy Research, University of Colorado, nin.tl/media-climate **2** Leo Barasi, *A climate paradox? The impact of severe weather events on concern about climate change*, unpublished MSc dissertation. **3** Neil T Gavin, 'Addressing climate change: a media perspective', *Environmental Politics*, September 2009. **4** Claire Saunders & Maria Grasso, *News prompts and Coverage of Climate Change in British Newspapers During Two Attention Cycles (1997-2004 and 2005-2009)*, working paper, forthcoming. **5** Maxwell T Boykoff, 'Flogging a Dead Norm? Newspaper Coverage of Anthropogenic Climate Change in the United States and United Kingdom from 2003 to 2006', *Area*, December 2007. **6** Maxwell T Boykoff & Maria Mansfield, ''Ye Olde Hot Aire': reporting on human contributions to climate change in the UK tabloid press', *Environmental Research Letters*, April 2008. **7** Saffron O'Neill, Hywel TP Williams, Tim Kurz, Bouke Wiersma & Maxwell Boykoff, 'Dominant frames in legacy and social media coverage of the IPCC Fifth Assessment Report', *Nature Climate Change*, March 2015. **8** Sapping Attention, nin.tl/screen-time-saps **9** Diego Román & KC Busch, 'Textbooks of doubt: using systemic functional analysis to explore the framing of climate change in middle-school science textbooks', *Environmental Education Research*, Vol 22, No 8, 2016. **10** For example, Dietram A Scheufele & David Tewksbury, 'Framing, Agenda Setting, and Priming: The Evolution of Three Media Effects Models', *Journal of Communication*, March 2007. **11** For example, Neil T Gavin & David Sanders, 'The Press and its Influence on British Political Attitudes under New Labour', *Political Studies Association*, October 2003. **12** John R Zaller, *The Nature and Origins of Mass Opinion*, Cambridge University Press, 1992. **13** In a 1998 paper, *Monica Lewinsky's Contribution to Political Science*, Zaller partly modified this argument, suggesting instead that public opinion is in fact strongly informed by real events, rather than only the view of media commentators. However, this was based on a study of Bill Clinton's approval rating and how it was affected by the strength of the economy, the crime rate and other visible factors closely related to most people's daily lives. When it comes to climate change, most people's experience continues to be through the media, and so the Receive-Accept-Sample model seems still to hold. **14** Leo Barasi, unpublished MSc dissertation, op cit. **15** Anthony Leiserowitz, 2009, op cit. **16** For example, Climate Outreach, nin.tl/breaking-climate-silence **17** London Gatwick, nin.tl/lgw-noise **18** Campaign for Better Transport & Fellow Travellers, nin.tl/runway-costs **19** Search of media database. **20** Kevin F Quigley, 'Bug reactions: Considering US government and UK government Y2K operations in light of media coverage and public opinion polls', *Health, Risk & Society*, Vol 7, No 3, 2005. **21** BBC, nin.tl/Y2000-bug **22** Ibid. **23** For example: Carbon Brief, nin.tl/alarming-misinterpretation; Carbon Brief, nin.tl/crop-yield-impact; Carbon Brief, nin.tl/ridley-misinterpreted **24** For example: Carbon Brief, nin.tl/ice-free-arctic; Carbon Brief, nin.tl/methane-emergency-claims **25** Pew Research Center, nin.tl/avoiding-the-news **26** Harvard Big Thinks, vimeo.com/10324258 **27** Richard E Zeebe, Andy Ridgwell & James C Zachos, 'Anthropogenic carbon release rate unprecedented during the past 66 million years', *Nature Geoscience*, April 2016. **28** Daniel Pauly, 'Anecdotes and the shifting baseline syndrome of fisheries', *Trends in Ecology & Evolution*, October 1995. George Monbiot also used the term in the same way in his excellent 2013 book, *Feral*. **29** Daniel Kahneman & Amos Tversky, 'Choices, Values, and Frames', *American Psychologist*, April 1984. **30** Daniel Kahneman, Jack L Knetsch & Richard H Thaler, 'Anomalies: The Endowment Effect, Loss Aversion, and Status Quo Bias', *Journal of Economic Perspectives*, Winter 1991. **31** This interpretation is suggested in Jeffrey J Rachlinksi, 'The Psychology of Global Climate Change', *University of Illinois Law Review*, 2000. **32** Charles G Lord, Lee Ross & Mark R Lepper, 'Biased Assimilation and Attitude Polarization: The Effects of Prior Theories on Subsequently Considered Evidence', *Journal of Personality and Social Psychology*, Vol 37, No 11, 1979. **33** George Marshall, 2014, op cit.

Chapter 6 – Nothing to worry about

1 House of Lords Debate, nin.tl/HoL-energy-bill **2** *Polar bear,* nin.tl/YT-polar-bear **3** *Don't give up*, nin.tl/YT-dont-give-up **4** Greenpeace, nin.tl/greenpeace-santa **5** George Lakoff, *Don't Think of an Elephant! Know Your Values and Frame the Debate*, Chelsea Green Publishing, 2004. **6** Although much of the likely increase in sea level is from the expansion of water as it warms, rather than from the increased water released by melting ice. **7** Yale University and George Mason University, nin.tl/six-americas-2011 **8** For example, *The Guardian*, nin.tl/4000-major-breaches **9** Various surveys in different English-speaking industrialized countries have found largely consistent proportions who report to be vegetarian over recent years. Collections of these surveys can be found at: Vegetarian Victoria, vegvic.org.au/statistics; Vegetarian Society, nin.tl/uk-vegetarians **10** *Vox*, nin.tl/less-meat; *The Guardian,* nin.tl/comparing-carnivores **11** Anthony Leiserowitz et al, 2009, op cit. **12** Donald Hine et al, op cit. **13** Anthony Leiserowitz et al, 2009, op cit. **14** Donald Hine et al, 2013, op cit. **15** A Spence, W Poortinga, C Butler & NF Pidgeon, 'Perceptions of climate change and willingness to save energy related to flood experience', *Nature Climate Change*, March 2011; Understanding Risk Research Group, nin.tl/flooding-perceptions **16** Edward W Maibach, Matthew Nisbet, Paula Baldwin, Karen Akerlof & Guoqing Diao, 'Reframing climate change as a public health issue: an exploratory study of public reactions', *BMC Public Health*, June 2010. **17** National Climate Change Adaptation Research Facility, nin.tl/climate-responses-aus **18** Google Trends, nin.tl/ebola-ferguson **19** World Health Organization, nin.tl/ebola-report **20** World Health Organization, nin.tl/malaria-report **21** Al Jazeera America, nin.tl/CAR-deaths **22** Leo Barasi, unpublished MSc dissertation, op cit. **23** Ipsos MORI, 2017, op cit. **24** Roy Morgan Research, nin.tl/economic-problems **25** Gallup, nin.tl/most-important-problem **26** IPCC, nin.tl/IPCC-comms-strategy **27** *The Independent*, nin.tl/4-degree-rise **28** Daniel Kahneman, *Thinking, Fast and Slow*, Farrar, Straus and Giroux, 2011. **29** Opinium, 2013b, op cit. **30** The question was asked in Celsius. **31** Michael J Benton & Richard J Twitchett, 'How to kill (almost) all life: the end-Permian extinction event', *Trends in Ecology & Evolution*, July 2003. **32** Shu-zhong Shen & Samuel A Bowring, 'The end-Permian mass extinction: a still unexplained catastrophe', *National Science Review*, December 2014. **33** John A Church, Peter U Clark et al, 'Sea Level Change', in Thomas F Stocker & Dahe Qin et al, 2013, op cit. **34** Opinium, 2013b, op cit. **35** Defra, nin.tl/defra-report **36** Ibid **37** Opinium, 2013a, op cit; Opinium, 2013b, op cit.

38 Kristie L Ebi & David Mills, 'Winter mortality in a warming climate: a reassessment', *WIREs Climate Change*, May/June 2013. **39** Jemma Gornall et al, 'Implications of climate change for agricultural productivity in the early twenty-first century', *Philosophical Transactions B*, September 2010.

Chapter 7 – Do you have to be leftwing to worry about climate change?

1 Naomi Klein, *This Changes Everything: Capitalism vs the Climate*, Allen Lane, 2014. **2** Climate Home, nin.tl/fossil-fuel-enemy **3** BBC, nin.tl/what-is-OECD **4** Margaret Thatcher Foundation, nin.tl/thatcher-speech **5** *The Washington Post*, nin.tl/pentagon-prep; Radio Australia, nin.tl/climate-military; *The Star*, nin.tl/operation-nanook; *The Guardian*, nin.tl/climate-conflict **6** NATO, nin.tl/stoltenberg-speech **7** IMF, nin.tl/IMF-climate **8** Goldman Sachs, nin.tl/goldman-climate **9** The Coca-Cola Company, nin.tl/coke-position **10** Grantham Research Institute on Climate Change and the Environment, nin.tl/domestic-dynamics **11** PSB, unpublished poll, 2016. **12** For example, John Rawls, *A Theory of Justice*, Harvard University Press, 1971. **13** David Harvey, *A Brief History of Neoliberalism*, Oxford University Press, 2005. It is sometimes argued that the term 'neoliberalism' is used to mean too many different things to be a useful concept. But with the definition given here, we have a specific meaning for it. For example, Taylor C Boas & Jordan Gans-Morse, 'Neoliberalism: From New Liberal Philosophy to Anti-Liberal Slogan', *Studies in Comparative International Development*, June 2009. **14** Oskar Lange, 'The role of planning in socialist economy', *Indian Economic Review*, August 1958. **15** Edmund Burke, *Reflections on the Revolution in France*, Oxford University Press, 1993. **16** A more recent example of this, from writers linked to a government on the Center-Left, was: Anthony Giddens & Patrick Diamond (eds), *The New Egalitarianism*, Polity Press, 2005. **17** *Mirror*, nin.tl/inequality-immigration; *The Telegraph*, nin.tl/left-behind-voters **18** For example, on climate change: YouGov, nin.tl/You-Gov-climate **19** To avoid adding further dimensions, these views are considered together, but openness to free trade can contrast with openness on social issues. For example, some free-market conservatives support free trade but are illiberal on social issues. In contrast, during his presidential primary campaign Bernie Sanders was open on many social issues but relatively closed on free trade (and so is shown as more closed than Hillary Clinton in the graphic). **20** For example, Michael Jacobs, 'Environmental Modernism: The New Labour Agenda', *The Fabian Society*, No 591, 1999 **21** Rachel Carson, *Silent Spring*, Houghton Mifflin, 1962. **22** Naomi Oreskes & Erik M Conway, 2010, op cit. **23** Gallup, nin.tl/fewer-enviros **24** Tim Jackson, *Prosperity Without Growth: Economics for a Finite Planet*, Earthscan, 2009. **25** Naomi Klein, 2014, op cit. **26** Ibid. **27** Ibid. **28** OpenDemocracy, nin.tl/climate-unspun **29** OpenDemocracy, nin.tl/media-covering **30** Gallup, 2016, op cit. **31** This was notably the case for the then-aspiring UK Prime Minister, David Cameron, who prominently travelled to the Arctic and changed his party's logo to be an oak tree. **32** Many studies have found this, for example: The Volker Alliance, nin.tl/global-trust; Political Studies Association, nin.tl/understanding-politics **33** The quality of the evidence about the economic costs and benefits means it is difficult to reach firm conclusions. It might be that models still underestimate the costs of inaction. For example, Nicolas Stern, 'Economics: Current climate models are grossly misleading', *Nature*, February 2016. **34** Noise of the Crowd, nin.tl/not-just-politicians **35** UK: Ipsos Mori, nin.tl/political-alignment; US: Gallup, nin.tl/conservatives-thread **36** Ibid. **37** 'Smug Alert!', *South Park*, Comedy Central, 29 March 2006. **38** In an ironic way, of course, given the problems of the association of climate change and the Left.

Chapter 8 – The pointy end

1 George RR Martin, *A Game of Thrones*, Voyager, 1996. **2** Terry Pratchett & Neil Gaiman, *Good Omens*, Victor Gollancz, 1990. **3** Connie Roser-Renouf, Edward Maibach, Anthony Leiserowitz & Seth Rosenthal, 2016, op cit. **4** Tom Crompton & Tim Kasser, *Meeting Environmental Challenges: The Role of Human Identity*, WWF-UK, 2009. **5** Crispin Tickell,

'Societal responses to the Anthropocene', *Philosophical Transactions A*, March 2011. **6** Leo Barasi, unpublished MSc dissertation, op cit. **7** Climate Outreach, nin.tl/narrative-workshops **8** See research cited in Chapter 4. **9** For example, Carbon Brief' nin.tl/media-reacts **10** Met Office, nin.tl/summer-2009 **11** BBC, nin.tl/sizzle-summer **12** House of Commons Debate, nin.tl/transport-disruption **13** 'The Uncertainty Handbook', University of Bristol, nin.tl/uncertainty-handbook **14** For example, *The Telegraph*, nin.tl/100-months; *The Independent*, nin.tl/six-months-left **15** H Damon Matthews & Andrew J Weaver, 'Committed climate warming', *Nature Geoscience*, Vol 3, 2010. **16** Steven J Davis, Ken Caldeira, H Damon Matthews, 'Future CO2 Emissions and Climate Change from Existing Energy Infrastructure', *Science*, Vol 329, No 5997, September 2010.

Chapter 9 – Tear down this wall

1 Climate Outreach, April 2014, op cit. **2** See Chapter 7. **3** *The New Statesman*, nin.tl/rightwing-thriving **4** IRENA, nin.tl/costs-fall **5** For example, *Fortune*, nin.tl/record-solar-price **6** Chris Goodall, *The Switch: How solar, storage and new tech means cheap power for all*, Profile Books, 2016. **7** For example, *Mark Lynas: How the planet can survive the age of humans*, Fourth Estate, 2011. **8** Opinium, 2013b, op cit. **9** With some justification – a proposed nuclear power station in the UK, at Hinkley Point, has an estimated cost that is more than four times that of the Large Hadron Collider in France and Switzerland. **10** *The Guardian*, nin.tl/climate-denialism **11** *An Ecomodernist Manifesto*, nin.tl/eco-mod-manifesto **12** *The Guardian*, nin.tl/eco-mod-critique **13** *The Guardian*, nin.tl/impressive-screw-up **14** Opinium, 2013a, op cit. **15** Common Cause Foundation, nin.tl/no-cause **16** Ibid. **17** Chris Rose, 2011, op cit.

Chapter 10 – Yes we can

1 House of Commons Debate, nin.tl/war-situation **2** 'The Gold Violin', *Mad Men*, AMC, 7 September 2008. **3** Garrett Hardin, 'The Tragedy of the Commons', *Science*, December 1968. Subsequent research has suggested that Hardin's theory is incomplete and that users of common resources often organize arrangements for their sustainable use. For example: David Feeny, Fikret Berkes, Bonnie J McCay & James M Acheson, 'The Tragedy of the Commons: Twenty-two years later', *Human Ecology*, March 1990. **4** *Keep America Beautiful*, nin.tl/littering-behavior **5** For example, in 2012 English councils issued nearly 64,000 litter fines, the equivalent of around 175 a day. This was presented as a large number when the figures were published, but it is the equivalent of one fine per 800 people. It wouldn't be hard to find more than 175 pieces of litter in a walk around a typical English city. BBC, nin.tl/litter-fine-cash-cow **6** Jos GJ Olivier et al, 2016, op cit. **7** China's emissions are around 30 per cent of the world total; the US's are around 14 per cent; the next largest is India's, which is just under seven per cent. Ibid. **8** CAIT Climate Data Explorer, nin.tl/CIAT-map **9** Alexa Spence & Nick Pidgeon, 'Framing and communicating climate change: The effects of distance and outcome frame manipulations', *Global Environmental Change*, October 2010. **10** Paul G Bain et al, 'Co-benefits of addressing climate change can motivate action around the world', *Nature Climate Change,* August 2015. **11** CAIT Climate Data Explorer, 2016, op cit. **12** UNFCCC, nin.tl/INDC-press-release **13** Carbon Brief, nin.tl/2027-peak **14** Climate Action Tracker, nin.tl/china-tracker **15** This cut is compared with 2005 levels; since emissions have already fallen since then due to reduced deforestation, this would suggest a small increase on 2012 levels. That would represent a large cut in the intensity of emissions from economic activity. Federative Republic of Brazil, nin.tl/brazil-contribution **16** Federal Democratic Republic of Ethiopia, nin.tl/ethiopia-contribution **17** IRENA, nin.tl/capacity-statistics **18** Jos GJ Olivier et al, 2016, op cit; Corinne Le Quéré et al, 2016, op cit. **19** I'm not sure anyone describes themselves as a bright-sider – the term seems only to be used as something to argue against. But many people who talk about climate change take an approach that can be labelled bright-siding, as described in David Spratt, 'Bright-siding climate advocacy and its consequences', *Climate Code Red*, April 2012, nin.tl/brightsiding **20** George Marshall, 2014, op cit. **21** Nicholas Stern, *The economics of climate change: the Stern review*, Cambridge University Press, 2007. **22** *The Guardian*, nin.tl/low-carbon-necessity **23** Institute for Health Metrics and Evaluation, nin.tl/ghdx-data **24** Some suggest that the quickest way to reduce indoor air pollution is to prioritize supplying cheap electricity wherever it is absent, even if this means burning more coal – but others argue that the lack of grid infrastructure in poorer countries means that it would still be cheaper to replace domestic cook stoves with ones powered by cleaner alternatives. **25** Paul G Bain et al, 2015, op cit, and personal communication with author. **26** Carbon dioxide-equivalent, to be precise. Peter Scarborough et al, 'Dietary greenhouse gas emissions of meat-eaters, fish-eaters, vegetarians and vegans in the UK', *Climatic Change*, July 2014. **27** There is no simple answer for what this figure should be. The one ton figure is from David JC MacKay, 2008, op cit. To check that it matches what we should expect, we can compare it with the figures we saw in Chapter 3 for the levels that global emissions need to fall by. The world will need to reduce gross emissions (before absorbing any greenhouse gases) to just a few billion tons of carbon dioxide a year if it is to avoid dangerous warming. So in a world with around eight or nine billion people, one ton per person seems a plausible – if generous – limit. **28** This is probably an underestimate as it counts only carbon-dioxide emissions and excludes other factors that increase the warming effect of emissions from flying. ICAO, nin.tl/carbon-emissions-calculator **29** Richard H Thaler, 'Anomalies: The Ultimatum Game', *The Journal of Economic Perspectives*, Autumn 1988. **30** *A Free Ride*, Fellow Travellers, 2017, afreeride.org **31** *The Guardian,* nin.tl/climate-march-2015 **32** Change.org, nin.tl/reinstate-clarkson **33** *The Guardian*, nin.tl/green-policies-axed **34** Committee on Climate Change, nin.tl/2015-progress-report